Darwin's Legacy

African Archaeology Series
Series Editor: Joseph O. Vogel

The **African Archaeology Series** is a series of volumes intended to present comprehensive and up-to-date summaries of current research on the African cultural past. The authors, who range from palaeoanthropologists to historical archaeologists, work in Africa and detail the results of their ongoing research interests. Though the essential subject matter of each volume is drawn from archaeology, they are equally dependent upon investigation of the historical and anthropological records as well. The authors understand the diversity and depth of African culture and history, and endeavor to explore the many sources of the African experience and the place of the African past and lifeways in the broader world. At the same time, the series permits Africanist scholars the opportunity to transform their field results into more general syntheses, giving context and meaning to a bare bones archaeological record, exploring and utilizing new techniques for explaining, as well as comprehending, the past.

Books in the Series
Chapurukha M. Kusimba, *The Rise and Fall of Swahili States* (1999)
Michael Bisson, S. Terry Childs, Philip de Barros, Augustin F. C. Holl, *Ancient African Metallurgy: The Socio-Cultural Context* (2000)
Innocent Pikirayi, *The Zimbabwe Culture: Origins and Decline of Southern Zambezian States* (2001)
Sibel Barut Kusimba, *African Foragers: Environment, Technology, Interactions* (2002)
Augustin F. C. Holl, *Saharan Rock Art: Archaeology of Tassilian Pastoralist Iconography* (2004)
J. D. Lewis-Williams, D. G. Pearce, *San Spirituality: Roots, Expression, and Social Consequences* (2004)
Peter Mitchell, *African Connections: Archaeological Perspectives on Africa and the Wider World* (2005)
Andrew B. Smith, *African Herders: Emergence of Pastoral Traditions* (2005)
Colleen E. Kriger, *Cloth in West African History* (2006)
Roger Blench, *Archaeology, Language, and the African Past* (2006)
Peter R. Schmidt, *Historical Archaeology in Africa: Representation, Social Memory, and Oral Traditions* (2006)
Pamela R. Willoughby, *The Evolution of Modern Humans in Africa: A Comprehensive Guide* (2006)
Sue Taylor Parker and Karin Enstam Jaffe, *Darwin's Legacy: Scenarios in Human Evolution* (2008)

Submission Guidelines
Prospective authors of single- or coauthored books and editors of anthologies should submit a letter of introduction, the manuscript, or a four- to ten-page proposal, a book outline, and a curriculum vitae. Please send your book manuscript or proposal packet to:

    African Archaeology Series
    AltaMira Press
    4501 Forbes Boulevard, Suite 200
    Lanham, MD 20706
    USA

Darwin's Legacy

Scenarios in Human Evolution

SUE TAYLOR PARKER
AND
KARIN ENSTAM JAFFE

ALTAMIRA
PRESS

A Division of
ROWMAN & LITTLEFIELD PUBLISHERS, INC.
Lanham • New York • Toronto • Plymouth, UK

ALTAMIRA PRESS
A division of Rowman & Littlefield Publishers, Inc.
A wholly owned subsidiary of The Rowman & Littlefield Publishing Group, Inc.
4501 Forbes Boulevard, Suite 200, Lanham, MD 20706
www.altamirapress.com

Estover Road, Plymouth PL6 7PY, United Kingdom

Copyright © 2008 by AltaMira Press

All rights reserved. No part of this publication may be reproduced, stored in a retrieval system, or transmitted in any form or by any means, electronic, mechanical, photocopying, recording, or otherwise, without the prior permission of the publisher.

British Library Cataloguing in Publication Information Available

Library of Congress Cataloging-in-Publication Data
Parker, Sue Taylor.
  Darwin's legacy : scenarios in human evolution / Sue Taylor Parker and Karin Enstam Jaffe.
    p. cm. — (African archaeology series)
  Includes bibliographical references and index.
  ISBN-13: 978-0-7591-0315-3 (cloth : alk. paper)
  ISBN-10: 0-7591-0315-1 (cloth : alk. paper)
  ISBN-13: 978-0-7591-0316-0 (pbk. : alk. paper)
  ISBN-10: 0-7591-0316-X (pbk. : alk. paper)
  1. Human evolution. 2. Human beings—Evolution. I. Jaffe, Karin Enstam, 1971– II. Title.
  GN281.P339 2008
  599.93'8—dc22

2007048162

Printed in the United States of America

∞™ The paper used in this publication meets the minimum requirements of American National Standard for Information Sciences—Permanence of Paper for Printed Library Materials, ANSI/NISO Z39.48-1992.

Contents

List of Sidebars	vii
Preface	ix
Acknowledgments	xi

PART I: Background

1	Scenarios in Human Evolution: Science or Fiction?	3
2	Guidelines for Modeling Human Evolution	13
3	Who Were Our Ancestors?	25

PART II: Origins Models

4	Origins of Human Subsistence	45
5	Origins of Bipedalism	67
6	Origins of Human Life History	83
7	Origins of Human Bodily Displays	105
8	Origins of Language	131
9	Origins of Human Mentality	155
10	Origins of Cultures	185

Epilogue: What about Bonobos?	205
References	209
Index	243
About the Authors	249

Sidebars

CHAPTER 1
Charles Darwin's Scenario for the Evolution of Bipedalism, Dexterous
    Hands, Smooth Skin, Language, and Higher Intelligence (1871)   6

CHAPTER 4
Bartholomew and Birdsell's Ecological Scenario for Hominid
    Subsistence (1953)   46
Dart's Killer Ape Scenario (1953)   50
Washburn and Lancaster's Hunting Scenario (1968)   54
Tanner and Zihlman's Scenario for Hominid Subsistence (1976)   56
Szalay's Scavenging Hypothesis (1976)   61
Wrangham, Jones, et al.'s Subsistence Scenario for Origins of
    Australopithecus and *Homo erectus* (1999)   62

CHAPTER 5
Gordon Hewes's Food Transport Scenario for Bipedalism (1961)   72
Roger Westcott's Upright Display Scenario (1967)   77
Richard Young's Throwing and Clubbing Scenario (2003)   80

CHAPTER 6
Lovejoy's Scenario for Origins of Human Life History, Subsistence, and
    Bipedalism (1981)   88

Lancaster and Lancaster's Parental Investment Scenario (1983)   93
O'Connell, Hawkes, et al.'s Grandmothering and the Evolution of *Homo erectus* (1999)   100

CHAPTER 7
Alister Hardy's Aquatic Scenario (1960)   111
Wheeler's Temperature Regulation Scenario (1984)   113
Desmond Morris's Naked Ape Scenario (1967)   118

CHAPTER 8
Hockett and Ascher's Revolutionary Language Origins Scenario (1964)   134
Gordon Hewes's Language Evolution Scenario (1973)   136
Armstrong, Stokoe, et al.'s Origins of Syntax Scenario (1994)   140
Robin Dunbar's Gossip Scenario for Language Evolution (1996)   144
Falk's Developmental Scenario for the Origin of Language (2004)   146
Noble and Davidson's Planning Scenario (1996)   150

CHAPTER 9
Parker and Gibson's Extractive Foraging Scenario for Evolution of Cognition (1979)   157
Wynn's Scenario for Spatial Intelligence (1989)   163
Mithen's Architectural Scenario for Mental Evolution (1996)   170
Wrangham and Peterson's "Demonic Male" Scenario for the Evolution of Violence and Intelligence (1996)   174
Geoffrey Miller's Sexual Selection Model for Hominid Brain Evolution (2000)   177

CHAPTER 10
Engels's Labor Scenario (1896)   187
Robert Bigelow's Warfare Scenario (1969)   192
Merlin Donald's Scenario for the Evolution of Culture, Language, and Cognition (1991)   196
Bingham's Complete Theory of Human Evolution (1999)   200

Preface

This book was inspired by a course I always wanted to teach, but never taught—a seminar on the history of ideas about the evolution of uniquely human characteristics. It was inspired by Darwin's book *The Descent of Man and Selection in Relation to Sex*. Previous books on the history of ideas about human evolution have focused on formulations of early-twentieth-century anthropologists who did not think in terms of Darwin's theory of natural selection (Bowler 1986; Landau 1991). In contrast, this book emphasizes the roles of natural and sexual selection in human evolution, revealing the remarkable prescience of Darwin's scenarios. In addition to bipedalism, life history, and facial and bodily displays, it discusses the evolution of language and culture.

Despite the four-field model of American anthropology, most courses on the history and theory of anthropology focus primarily on cultural anthropology, reflecting the traditional dominance of that field. My coauthor and I hope our book will find its way into some of these courses. Perhaps someday she will teach the seminar I never taught.

—Sue Taylor Parker

Acknowledgments

We would like to thank Joe Vogel for inviting and nurturing this book, Jack Meinhardt for editorial guidance, and Janice Braunstein for overseeing its production. We also want to thank our colleagues, teachers, and students for stimulating our interest in the history of anthropology. Finally, we want to thank our spouses for their support in this project.

I
BACKGROUND

1

Scenarios in Human Evolution: Science or Fiction?

It has often and confidently been asserted, that man's origin can never be known; but ignorance more frequently begets confidence than does knowledge. It is those who know little, and not those who know much, who so positively assert that this or that problem will never be solved by science. The conclusion that man is the co-descendant with other species of some ancient, lower, and extinct form, is not in any degree new.

—*Darwin 1871, 290*

What makes us human? When did our unique characteristics appear? What caused them to appear? These questions have fascinated laymen and scholars, both ancient and modern. Since Darwin's time at least, anthropologists have proposed a series of scenarios to explain the origins of uniquely human characteristics.

Descriptions of unique features of humans, including bipedalism, language, and increased cranial capacity, go back as far as the fourth or fifth century BC (Stoczkowski 2002). For the most part, these ancient prescriptions agree with more recent formulations. From Darwin's time on, most investigators have agreed that humans differ from other primates in the following characteristics: bipedal walking, greater manual dexterity and worked stone tools, brain enlargement (encephalization), prolonged childhood and juvenile life history stages, language (especially speech production), and culture. They have disagreed, however, on where, when, how, and why these characteristics emerged.

Some themes are apparent. Most of the scenarios proposed to explain bipedal walking, canine tooth reduction, manual dexterity, and tool use are based on subsistence activities, for example, hunting, scavenging, gathering, excavating, or extractive foraging. In contrast, one scenario for bipedalism is based on physiological functions of temperature regulation. Still others focus on social activities such as defense, display, food sharing, and caretaking. Some are based on analogies with savanna-living primates or carnivores, some are based on homologies with our closest living relatives (chimpanzees), others on analogies with human hunter-gatherers.

Whether they disparage scenarios of human evolution as "just so stories" or elevate them to the status of scientific hypotheses, people are fascinated by stories about the history of our species. In hunter-gatherer preliterate societies, these narratives take the form of what anthropologists call origin myths. In early civilizations, they sometimes took the form of stories about powerful, creative gods whose lineages culminated in the house of the ruler. Beginning with Darwin, these narratives have taken the form of evolutionary reconstructions (P. Bowler 1986).

This book focuses on about thirty adaptive scenarios proposed after Charles Darwin (1859, 1871) wrote *On the Origin of Species* and *The Descent of Man*. These views represent a significant sampling of the many scenarios proposed over the last century. We chose them for their historical significance and their influence on anthropology textbooks and fellow scholars, as well as for their intrinsic interest. We also include some improbable scenarios, like the aquatic scenario, which have stimulated continuing debate and/or elaboration. We try to place these scenarios in the context of changing knowledge about human fossils and primate behavior, as well as changing theoretical emphases.

We organize discussion of these scenarios topically by characteristics, including bipedalism, subsistence patterns, life history, bodily displays, language, intelligence, and culture. We discuss scenarios that cover more than one topic in more than one chapter. Short summaries of selected scenarios appear as sidebars in each topical chapter. This chapter features Darwin's scenario, the first evolutionary scenario and also one of the most comprehensive.

SCIENCE OR MYTHOLOGY?

As indicated above, scenarios of hominin origins have a questionable reputation in some quarters. For example, in his historical study, Wictor Stoczkowski argues that scenarios of human evolution have remained static over many centuries. According to him, "the core of the knowledge employed in humanization scenarios is based on a relatively inert structure that prolongs the tradition of conjectural anthropology and has not been remodeled, in its broad outlines, either under the influence of new paleontological and archaeological discoveries or as a consequence of the theoretical developments in biology" (Stoczkowski 2002, 127). He also charges that these scenarios are largely untestable.

Likewise, in her literary analysis, M. Landau (1991) argues that late-nineteenth- and early-twentieth-century scenarios of human evolution share a common narrative structure of heroic transformation. In each case she follows the hero from an initial situation through a change followed by a departure and a test or a series of tests. She notes that a donor, which can be natural forces, "bestow[s]on the hero gifts—intelligence, tools, a moral sense—that transform him into a primitive human" (Landau 1991, 11).

In contrast, we believe that the narrative form arises as much from the nature of evolution as from the nature of storytelling. Evolution implies history and history implies narrative. Evolution is a historical process that depends in part on chance and in part on complex unpredictable contingencies like climate change. This means that the history of a species is unique and unrepeatable. The exact history of a given species and ecosystem is never exactly repeated. History implies narrative, and evolution is history writ large (Gould 2002).

According to our contemporary understanding, evolution occurs through a variety of processes, including the generation of variants through mutation and sexual reproduction and differential survival and reproductive success of some variants as compared to others, either by chance or through selection (Mayr and Provine 1989). Only selection results in adaptations. Because forms of organic life have adapted to changing ecosystems through historical processes of natural selection, their adaptations have unique histories. Again evolutionary history implies narrative, specifically, narratives about adaptation. Before addressing the question of how adaptive histories are reconstructed, we should note some differences between evolutionary narratives and other narrative forms.

The nature of evolution decrees that adaptive scenarios differ from origin myths and heroic transformations, Greek and Shakespearean plays, and other

Charles Darwin's Scenario for the Evolution of Bipedalism, Dexterous Hands, Smooth Skin, Language, and Higher Intelligence

Charles Darwin (1871, 442) tells the reader that he has two objects in writing *The Descent of Man:* "Firstly to shew that species had not been separately created, and secondly, that natural selection had been the chief agent of change." He divides his discussion of human evolution into two sections, the first focused on bodily structures involved in locomotion and prehension and the second on intellectual, social, and moral facilities.

He begins his discussion of bipedalism remarking that

> Man alone has become a biped; and we can, I think, partly see how he has come to assume his erect attitude, which forms one of his most conspicuous characters. Man could not have attained his present dominant position in the world without the use of his hands, which are so admirably suited to act in obedience to his will. . . . But the hands and arms could hardly have become perfect enough to have manufactured weapons, or to have hurled stones and spears with true aim, as long as they were habitually used for locomotion and for supporting the whole weight of the body. (Darwin 1871, 434)

He continues by suggesting how natural selection would have favored freeing of the hands: "They would thus have been better able to defend themselves with stones or clubs, to attack their prey, or otherwise obtain food" (434). Previously Darwin had noted the manual dexterity required for tool use:

> Even to hammer with precision is no easy matter. . . . To throw a stone with as true an aim as a Fuegian . . . requires the most consummate perfection in the correlated action of the muscles of the hand, arm, and shoulder, and, further, a fine sense of touch. In throwing a stone or spear, and in many other actions, a man must stand firmly on his feet; and this again demands the perfect co-adaptation of numerous muscles. To chip a flint into the rudest tool, or to form a barbed spear or hook from a bone, demands the use of a perfect hand. (Darwin 1871, 432)

Invoking the idea of feedback, Darwin (1871, 435) concludes that "the free use of the arms and hand, partly the cause and partly the result of man's erect position, appears to have led in an indirect manner to the modifications of structure. The early male forebearers were, as previously stated, probably furnished with great canine teeth; but as they gradually acquired the habit of using stones, clubs, or other weapons for fighting with their enemies or rivals, they would use their jaws and teeth less and less." He follows this with a reference to the parallel case of loss of canine teeth in male ruminants when they acquired horns.

Darwin remarks that naked skin is another conspicuous difference between man and "lower animals." He attributes this feature to sexual selection for ornamental features: "The females apparently first had their bodies denuded of hair, also a sexual ornament. . . . It is not improbable that the females were modified in other respects for the same purpose and by the same means; so that women have acquired sweeter voices and become more beautiful than men" (1871, 907). Finally, in discussing the enlarged brain of man, he notes that this is generally attributed to higher intellectual powers.

Regarding these powers, he endeavors to show the continuity between nonhuman and human mental abilities by reviewing the abilities of many animal species, including monkeys and apes and elephants:

> It has, I think, now been shewn that man and the higher animals, especially the Primates, . . . have the same senses, intuitions, and sensations—similar passions, affections, and emotions, even the more complex ones, such as jealousy, suspicion, emulation, gratitude, and magnanimity; they practice deceit and are revengeful; they are sometimes susceptible to ridicule, and even have a sense of humor; they feel wonder and curiosity; they possess the same faculties of imitation, attention, deliberation, choice, memory, imagination, the association of ideas, and reason, though in very different degrees. (456)

Based on his knowledge of primitive tribes and of human history, Darwin (1871, 874) concludes that man's mental abilities plus "genius" defined by energy, patience, and perseverance, arose "partly through sexual selection—that is, through the context of rival males, and partly through natural selection—that is, from success in the general struggle for life, and as in both cases the struggle will have been during maturity, the characteristics will have been transmitted more fully to the male than to the female offspring." He goes on to note that, like other secondary sexual characteristics, higher mental facilities are transformed at puberty.

In his discussion of man's unique capacity for language, Darwin suggests that this ability arose through sexual selection on males for singing during courtship:

> I cannot doubt that language owes its origin to the imitation and modification of various natural sounds, the voices of other animals, and man's own instinctive cries, aided by signs and gestures. When we treat of sexual selection, we shall see that primeval man, or rather some early progenitor of man, probably first used his voice in producing true musical cadences, that is, in singing, as do some of the gibbon-apes at the present day; and we may conclude from a widely spread analogy, that this power would have been especially exerted during the courtship of the sexes—would have expressed various emotions, such as love, jealousy, triumph—and would have served as a challenge to rivals. (1871, 463)

Darwin also discusses the evolution of man's moral sense, suggesting that it was inevitable in a social species with developed intellect.

kinds of traditional stories. First, they lack an oracle that foretells the unfolding of future events. There can be no oracle because there is no plot, no larger purpose or plan, to predict the future course of evolution. They also lack a purposeful hero who struggles to bring about a given end, or whose tragic character flaw dooms him to a tragic end. There can be no purposeful protagonist because evolution has no purpose. There can be no tragic flaw because evolution lacks any internal guiding force causing organisms to evolve toward a given end. Because we are looking backward, of course, we know the outcome of the story. This retrospective telling may lead the storyteller and his audience astray by couching the story in these mythic terms.

What kind of narratives lack plots, characters, and heroes? How are they constructed? What are the rules? What are the standards for judging the relative merits of one scenario as compared to another? At a minimum, these narratives should be consistent with fossil and archaeological data, with knowledge of the anatomy and behavior of closely related living species, and with modern understanding of evolutionary processes.

Like other scientific hypotheses, these scenarios should explain how things happened, with the proviso that there are at least four complementary kinds of causal explanations (Tinbergen 1963), two of which we will discuss here. *Proximate explanations* focus on how things work, that is, the physiological, psychological, and physical mechanisms underpinning a behavior or structure (it can also include details of their development). *Ultimate explanations* focus on how things got to be that way, that is, the evolutionary history of a lineage and the selective forces that favored these proximate mechanisms. A complete picture includes both kinds of explanation.

As we shall see, most adaptive scenarios focus on how things got that way. Many hominization scenarios are based on comparisons with other primate species and/or with living hunter-gatherers. Some anthropologists concerned with guidelines for creating models of human evolution distinguish between referential models and theoretical or conceptual models. Whereas referential models use one living species as the referent, conceptual models allow "strategic modeling" based on theories of natural and/or sexual selection, kin selection, or optimal foraging (Tooby and DeVore 1987). Referential species could show either shared adaptations or independently evolved (convergent) adaptations. In any case, the two kinds of models should be complementary.

The topical chapters evaluate selected "hominization scenarios" to what degree they are consistent with existing data and theory. As a framework for that analysis (see the next chapter for details), we propose that explanatory narratives of human adaptation should identify the following components:

1. the evolving lineage and its closest relatives;
2. the time and location in which they lived;
3. features of the environments in which this lineage lived;
4. key adaptive modifications, that is, new adaptations they displayed;
5. proximate mechanisms and functions of these new adaptations;
6. precursors of these new adaptations (perhaps in ancestral species);
7. the nature of selective forces operating on these adaptations;
8. evolutionary sequence of adaptations;
9. consistency with paleontological and archaeological data;
10. consistency with primatological and ethnological data.

Whenever possible, we begin each topical chapter with a brief description of the relevant characteristics of the fossils of each stage of human evolution, beginning with the last common ancestor of humans and chimpanzees and ending with modern humans. These are based on a variety of sources including studies of modern humans and chimpanzees, fossils, and archaeological data.

It is important to emphasize that knowledge of hominin evolution has expanded greatly since Darwin's time, when only a few hominin fossils were known. The fossil record includes thousands of specimens and scores of species dated between six million and twenty-five thousand years ago. Neanderthals were discovered in the 1850s, *Homo erectus* in the 1890s, and australopithecines, in the 1920s. From the 1970s through the 1990s paleoanthropologists discovered three- to five-million-year-old australopithecines characterized by small brains, short legs and long arms, and an absence of stone tools.

These unexpected discoveries caused anthropologists to reexamine earlier scenarios based on the assumption that the earliest hominins were essentially like modern hunter-gatherers. Likewise, between 1992 and 2002, discoveries of potential hominin precursors to australopithecines and analyses of the habitats they occupied have led to a reexamination of earlier scenarios based on the assumption that "hominization" occurred when our earliest ancestors moved out into the open savanna. It now appears that the earliest hominins lived and evolved in a forested habitat (e.g., Wolde-Gabriel, Haile-Selassie, et al. 2001), requiring a modification of scenarios that rely on savanna habitat as the impetus for the evolution of bipedalism.

In the years since Darwin, expanding fossil and molecular data have clarified the location and span of hominin evolution. They reveal that the earliest hominins lived as long as five or six million years ago in Africa. Recent discoveries and redating of earlier discoveries reveal that *Homo erectus* was in eastern Europe (Gabunia, Vekua, et al. 2000) and Southeast Asia as early as 1.8 million years ago (Swisher, Curtis, et al. 1994; Swisher, Rink, et al. 1996) and into China as early as a million years ago (e.g., Zhisheng and Kun 1989; Etler 1996). In addition to this early migration by *Homo erectus*, current evidence suggests at least two other major migrations out of Africa. One was of archaic *Homo sapiens* in continuing waves of migration into Asia and through Israel into Europe beginning approximately eight hundred thousand years ago (e.g., Ascenzi, Biddittu, et al. 1996; Manzi, Mallegni, et al. 2001). The other migration was of anatomically modern *H. sapiens* into the Middle East and Asia (sixty thousand years ago), Australia (forty thousand years ago), and Europe (forty thousand years ago) (Wells 2002).

Growing fossil and archaeological data have clarified the sequence of evolution of uniquely hominin characteristics. They have revealed, for example, that changes in locomotion and dentition preceded worked stone tools and changes in brain size. These data suggest that each major innovation evolved in two or three major steps. Early bipedal australopithecines, for example, retained relatively long arms and feet with slightly divergent great toes, which apparently allowed efficient tree climbing. These locomotor patterns coincided with reduced but still prominent canine teeth. Beginning about 1.8 million years ago, *H. erectus* evolved longer legs and shorter arms and feet without divergent toes, which apparently allowed efficient striding and running. A more modern thorax and diaphragm allowing more efficient breathing accompanied this remodeling. Increased brain size coincided with these changes.

The earliest known worked stone tools, Gona or Oldowan choppers (Semaw, Renne et al. 1997), appeared at about 2.5 million years ago associated with late australopithecines and early *Homo*. Archaeological discoveries of the earliest stone tools associated with animal bones (cracked to expose the marrow cavity), combined with taphonomic reconstructions of scavenging behavior in dry woodlands, have confirmed the dietary potential of the scavenging subsistence strategy.

In addition to fossil and archaeological data, studies of primate behavior and ecology have also influenced hominization scenarios. In the 1950s and 1960s, studies of baboon ecology stimulated scenario building, as did new revelations about chimpanzee behavior in the 1980s and 1990s. Increased field data and more sophisticated techniques for analyzing behavioral evolution have resulted in a new emphasis on chimpanzee behaviors as models for the behavior of the last common ancestor of chimpanzees and humans.

Long-term field studies of several populations of chimpanzees have brought new insights into the variation and complexity of their activities, particularly their hunting behavior, their territoriality and raiding, and their culturally variable tool use (e.g., Goodall 1986; Boesch and Boesch-Acherman 2000). Likewise, studies of symbol learning by captive chimpanzees have informed discussions of language evolution (e.g., Hewes 1973; D. Armstrong, Stokoe et al. 1994). Sociological factors, notably questioning of male dominance in paleoanthropology and the rise of feminism, influenced scenarios in the 1960s and 1970s (Tanner and Zihlman 1976; Zihlman and Tanner 1979; Fedigan 1982). In the 1990s sexual selection theory and kin selection theory stimulated new scenarios about bipedalism and subsistence (Wrangham and Peterson 1996).

All of these advances in knowledge provide a golden opportunity to reevaluate classic scenarios of the origins of key hominin characteristics. We approach this first by discussing concepts of adaptation and methods for identifying adaptation; second, by outlining recent hominin fossil data; and third, by analyzing influential scenarios using the framework set out above.

In contrast to Stoczkowski (2002), we conclude that genetics, fossils, archaeological data, and primate behavior studies have helped answer such questions as where, when, and sometimes how hominin evolution occurred. They can tell us which species first displayed which characteristics, the sequence in which these characteristics evolved, in which ecological contexts they evolved, and, sometimes, what adaptive functions these characteristics served. In the end, however, it is interesting to note how well Darwin's scenario has stood up in light of modern techniques and fossil and natural history data (Milner 1994).

2

Guidelines for Modeling Human Evolution

The fact, as we have seen, that all past and present organic beings can be arranged within a few great classes, in groups subordinate to groups, and with extinct groups often falling in between the recent groups, is intelligible on the theory of natural selection with its contingencies of extinction and divergence of characters. On these same principles we see how it is, that the mutual affinities of the forms within each class are so complex and circuitous. We see why certain characters are far more serviceable than others for classification. . . . The real affinities of all organic beings, in contradistinction to their adaptive resemblances, are due to inheritance or community of descent.

—*Darwin 1871, 366*

Anthropologists have a long history of constructing scenarios about the evolution of human characteristics (P. Bowler 1986; Landau 1991; Stoczkowski 2002). As we noted in chapter 1, these scenarios have used a variety of approaches to explain human traits. Much of this modeling has occurred without the benefit of clear-cut guidelines or methods. The wild and woolly nature of some scenarios has led at least two authors to focus on the mythic structure of these scenarios (i.e., Landau 1991; Stoczkowski 2002). In contrast, several anthropologists have examined the role that paleontological, archaeological, and ethnographic data have played in scenarios of human evolution (Isaac 1980; Fedigan 1982; M. Bowler, Pilbeam, et al. 1982).

Like most anthropologists, we believe these scenarios are largely testable. Recent approaches to evolutionary reconstruction provide us with tools for judging their plausibility. In some cases, these approaches allow us to figure out in whom, when, how, and why a given character arose. Before discussing these approaches, we need to explore the many meanings of the concept of adaptation, because whichever technique they use, most scenarios attempt to explain the evolutionary advantage of one or more hominin characters.

ADAPTATIONS, EXAPTATIONS, AND APTATIONS

Adaptations

According to modern evolutionary theory, *adaptations* are characteristics designed for a particular function through natural selection. They arise from mutant forms of preexisting characteristics and spread through a population when they give their bearers an advantage in survival and reproduction. It is important to recognize that not all beneficial characteristics are adaptations. Since Darwin's time biologists have known that some characteristics are beneficial side effects of adaptations. Driving, for example, is a beneficial side effect of our hand-eye coordination and fine motor skills (Williams 1966). Likewise, writing is a side effect of these same skills combined with language. Moreover, some features are merely correlated with adaptations, while others are selectively neutral and still other, deleterious, traits may have spread through chance.

Exaptations

In recent years, some biologists have argued that the term *adaptation* should be restricted to features selected for their current use, and the term *exaptation* should be used to refer to features "co-opted" for their current use after originating for some other use; they recommend the term *aptation* as a general term referring to characters that increase fitness, whichever way they arose (Gould and Vrba 1982; Gould 2002). These biologists prefer the term *exaptation* to the term *preadaptation*, which previously served the function of referring to preexisting traits that have become useful in a new context.

It is important to note that when exaptations stimulate new rounds of selection that refine their new functions, they become adaptations. Exaptations may become refined through selection of associated characteristics that render them especially efficient, or fast, or enduring, thereby leading to a new adaptation. Convergent utility of several preexisting adaptations (exaptations) is another important phenomenon to keep in mind in phylogenetic reconstruction and scenario building (Gould 2002). This is important because new adaptations ("aptations") can arise from two or more preexisting aptations. Bipedal walking, for example, may have been favored for aggressive displays, throwing, and tool use before being selected for object transport.

Identifying Adaptations

Biologists have spent considerable effort establishing criteria for recognizing adaptations. The most definitive approach is to demonstrate increased survival or reproductive success associated with the presence of the feature (Williams 1992). This works best when there is trait variation within a population. In some cases, as for example, in demonstrating that people with sickle cell trait are more resistant to malaria and survive and reproduce better than those without it, biologists have documented selection in action (Livingstone 1958).

When characteristics are already present in all members of a species, the presence of adaptations can be inferred through a variety of methods. The classic approach is the "argument by design" wherein biologists argue for a function by analogy with design for similar function by human engineers. Dawkins (1986), for example, compares the sonar of bats to mechanical sonar devices designed by naval engineers.

Another method is to see whether the same feature is present in all the species in a genus or family, that is, among a closely related group of species with a recent common ancestor (Coddington 1988). If, for example, all members of a lineage share a trait (e.g., the brachiation complex of hominoids), it may be inferred that the trait is an adaptation inherited from the common ancestor. The other possibility is that it evolved in parallel in two or more closely related species (e.g., prehensile tails in various New World monkeys) owing to exploitation of similar ecological opportunities, rather than inheritance from a common ancestor.

PHYLOGENETICS

In comparing similar organs and behaviors, biologists have long distinguished *homologous characteristics* in closely related species (e.g., the front limbs of bats, moles, and humans) that were inherited from a common ancestor from *analogous characteristics* (e.g., wings in bats and birds) that arose independently in distantly related species owing to similar selection pressures.

Phylogeneticists (Hennig 1979) added a new twist, distinguishing two kinds of homologous characteristics used to compare species. They realized that homologous characteristics shared by all the species in a lineage are not very helpful for sorting out who within that group is more closely related to whom. To do that you must look outside that group to find species that lack that characteristic (the out-group). Or, you need to find a derived characteristic shared by some species in your in-group but not others.

Categorizing Traits According to Their Evolutionary Relationships

Phylogenetic taxonomists distinguish five categories of relationship among traits (table 2.1) based on the sequence in which traits arose and who in a given lineage displays them. The two most basic categories are *ancestral* and *derived* traits. Ancestral traits can be *shared*. Derived traits can be *shared* or *unique* among species. There are two important points to remember about the categories listed in table 2.1. First, only *shared derived characters* are useful for determining which groups are more closely related to one another. Second, the category of a specific trait (table 2.1) depends on its pattern of occurrence or nonoccurrence among the species being compared. We can illustrate both of these points with one example.

Let's say we want to compare a monkey, a lemur, and a hedgehog to determine which two of the three are most closely related. Let's also say we know nothing else about these three species that will help us determine their relatedness. If several similar-looking species display a trait in common, evolu-

Table 2.1. Five Categories Evolutionary Biologists Use to Classify Characteristics

Category*	Definition
Ancestral trait (plesiomorphy)	Trait that evolved early in a lineage
Derived trait (apomorphy)	Trait that evolved late in a lineage
Shared ancestral trait (symplesiomorphy)	Trait that evolved early in a lineage and is shared by all members of that lineage
Shared derived trait (synapomorphy)	Trait that evolved late in a lineage and is only displayed by members of the lineage that evolved after the trait appeared
Uniquely derived trait (autapomorphy)	Trait that evolved late in a lineage but is only displayed by one member of the lineage

*The category of a trait depends on which species are being compared.

tionary biologists generally assume the trait evolved once in a common ancestor and was passed down to all of the descendants. It is more likely that a novel trait evolved once and was passed on to descendant species than that it evolved independently two or more times. (If the two species differ in many other characters, their isolated similarities probably arose through convergent, or parallel, evolution.)

If we want to determine which of these species (monkey, lemur, hedgehog) are most closely related, we look for traits shared by two of the three groups signaling that the trait arose after one of the species branched off the lineage, but before the other two diverged. If we can do this, we can determine which of the three are most closely related by virtue of shared derived characters.

Traits shared by all three species are no help. The presence of placental birth, for example, cannot help us decide how closely related monkeys, lemurs, and hedgehogs are because all three share this character. In other words, the placenta is a *shared ancestral character* (table 2.1) in our in-group. Because placentas evolved early in the history of mammals, this trait does not help us evaluate how closely related these species are. This addresses the first issue of taxonomic relevance—not all traits are equally useful in helping distinguish how groups are related.

If we want to determine relatedness among these three groups, we need to choose a different trait. We might notice, for example, that monkeys and lemurs share fingernails but that hedgehogs do not. In this case, fingernails are a shared derived character (table 2.1) of monkeys and lemurs, and indicate that these two species share a more common recent ancestor (which also had fingernails) than either shares with the hedgehog. Thus, we infer that the ancestor of hedgehogs branched off from the line leading to monkeys and lemurs before the appearance of fingernails, and monkeys and lemurs diverged sometime after fingernails appeared. Therefore monkeys and lemurs are more closely related to one another than either is to the hedgehog.

But the placenta can help us determine relatedness among other species. Again, the placenta is not taxonomically relevant when comparing placental mammals because they all share the trait. If, however, we add kangaroos to our species comparison, the placenta becomes a useful trait. It tells us that our in-group members of placental mammals are more closely related to each other than to kangaroos, which lack this trait. We can infer that placental birth, which distinguishes these two groups, must have arisen in the common ancestor of monkeys, lemurs, and hedgehogs after they separated from kangaroos. By adding kangaroos to the comparison, we have changed the category of our original trait (the placenta) from a shared ancestral trait to a shared derived trait (see table 2.1).

It bears repeating that the category of a trait, and therefore its taxonomic relevance for determining relatedness among species, depends on which species are being compared. Within placental mammals, for example, the placenta is an ancestral trait, shared by all. But when we compare a group of placental mammals with a nonplacental mammal, the placenta becomes a shared derived trait because it arose in mammals after the marsupials branched off.

Family Trees

So, all mammals that have a placenta share a common ancestor. They are more closely related to one another than they are to nonplacental mammals, which branched off before the appearance of the placenta. Based on both fingernails and the placenta, we can determine which groups of species (i.e., lemurs, monkeys, hedgehogs, and kangaroos) branched off earliest (figure 2.1). Family trees of different groups of organisms can be built up this way.

As we shall see in chapter 3, family trees arrange species in closely related groups according to how recently they shared a common ancestor. In recent years, however, taxonomists have used molecules, proteins, and DNA, rather than anatomical and behavioral characteristics, as the basis for constructing family trees of primates (Sarich 1968; Ghiselin 1991; Ruvolo 1994).

In the example above, we have determined the relationship among four species. Based on our knowledge, we have placed these species in a *cladogram*, or a tree, according to the number of derived features they share. The more derived traits species share, the more closely related they are (i.e., their branches connect to each other before connecting to other branches in the cladogram). Our cladogram in figure 2.1 indicates that lemurs and monkeys share a more recent common ancestor (they are both primates) because they share a derived trait (fingernails) that none of the other species in the cladogram display.

Reconstructing the Evolution of Characters

These same concepts can be used to reconstruct the evolution of characteristics. There are several steps in this approach to evolutionary reconstruc-

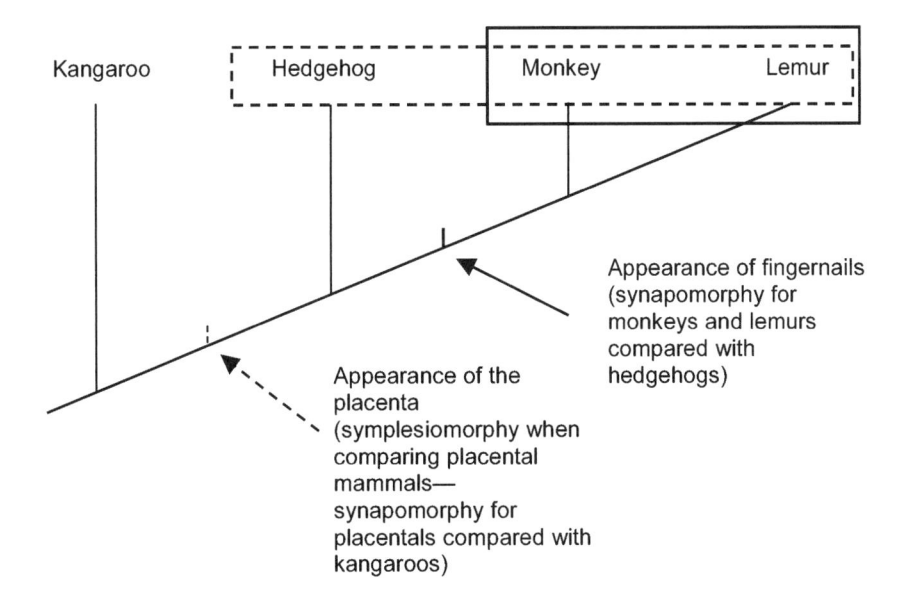

FIGURE 2.1.
Cladogram showing how two traits (the placenta and fingernails) can be used to determine which species branched off earlier and which later. The hedgehog, monkey, and lemur all display a placenta (dashed box), which makes the placenta a shared derived trait when comparing these three species to the kangaroo, a marsupial. But the placenta cannot help determine relatedness *within* the group because they all display it (it is a shared ancestral trait). We use a different trait (fingernails) to distinguish between the hedgehog, monkey, and lemur. Only monkeys and lemurs have fingernails (solid box), which makes fingernails a shared derived trait in monkeys and lemurs. So monkeys and lemurs are most closely related, followed by the hedgehog, and lastly, the kangaroo.

tion. The first step involves comparing our focal characteristics in closely related and distantly related species in order to identify shared ancestral characters, shared derived characters, and uniquely derived characters. The second step involves finding an independently derived family tree based on other characteristics—these days family trees based on DNA data are commonly used. (In order to avoid circularity, however, the family tree must be based on different characteristics than those whose evolution we are tracing.) The third step involves mapping these characteristics onto the independently derived family tree. Keep in mind that mapping characteristics onto phylogenies is an attempt to reconstruct the evolution of those characters.

Let us look at the distribution of two kinds of locomotion, brachiation and knuckle walking, in a branching diagram (or cladogram) of Old World monkeys, lesser apes, and great apes to illustrate this (figure 2.2). First, we map the character states for both brachiation and knuckle walking. We know that brachiation, or arm-over-arm suspensory locomotion under tree branches, occurs in gibbons and siamangs (the lesser apes) as well as in orangutans, gorillas, and chimpanzees (the great apes), but not in Old World monkeys.

Knuckle walking, in contrast, occurs in African apes but not in lesser apes or Old World monkeys (orangutans display a form of terrestrial locomotion known as fist walking). It does not occur in modern humans. If it occurred in hominins, we could say that knuckle walking is a shared derived trait that arose in the common ancestor of gorillas, bonobos, and chimpanzees. But given its absence in modern humans, we can't be sure. Looking for signs of knuckle walking in early hominins can help decide whether it evolved independently in gorillas and chimpanzees (see chapter 5). Certainly, brachiation is a shared ancestral trait among great apes because it arose in the common ancestor of all apes, before the lesser apes (gibbons and siamangs) branched off. (See chapter 5 for drawings of these locomotor patterns. Also, see figure 2.2. Note that branching points in the trees, or nodes, indicate a speciation event.)

The dates of divergence of African apes from orangutans (based on independent data) tell us the shared derived knuckle-walking trait may have arisen about nine million years ago (figure 2.2). To reconstruct the evolution of specific characters (i.e., brachiation and knuckle walking), we need to employ information about the form and function of the characters. Data from functional morphology tell us that knuckle walking is a form of terrestrial locomotion derived from the long arms and shortened legs involved in the shared ancestral trait of all apes, arboreal brachiation.

When anthropologists speak of uniquely human characteristics, they usually mean that these traits are *uniquely derived* when compared to traits of living African apes. When we say, for example, that bipedal locomotion is unique to humans, we are implicitly comparing humans to living African apes. On the

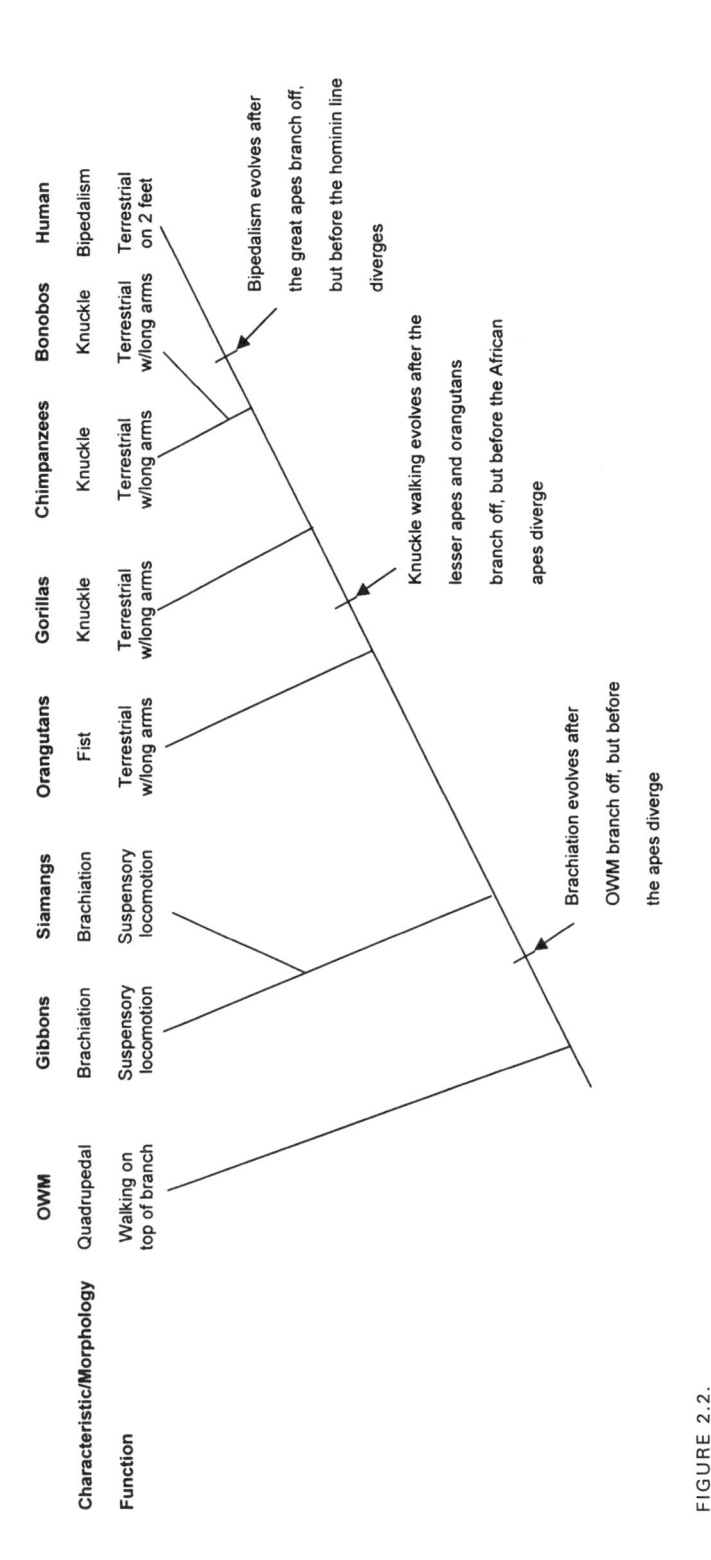

FIGURE 2.2.
Cladogram showing character reconstruction of brachiation and knuckle walking among the lesser and great apes (with Old World monkeys [OWM] as the out-group).
Source: Adapted from Coddington, J. A. 1988. Cladistic tests of adaptational hypotheses. *Cladistics* 4:3–22.

other hand, if we compare humans to most fossil hominins, bipedalism is a shared derived trait. In this case the fossil evidence is clear. As we will see, in other cases, such as language, the fossils are silent.

Selecting taxonomically relevant characteristics is an important issue in evolutionary reconstruction (Clark 1964). Shared ancestral and shared derived characters are usually developmentally complex and evolutionarily stable through adaptive radiations or speciation events. Brachiation, for example, is behavior made possible by a functional complex of bones, muscles, glands, and nerves. Knuckle walking is another. Placental birth is yet another.

In contrast, some characters are unstable and rapidly evolving. This is true, for example, of genitals and courtship patterns. These traits often change through sexual selection during speciation (West-Eberhard 1983; see chapter 7). Only biologists familiar with the anatomy and behavior of related species are in a position to identify meaningful functional complexes.

Phylogenetic Criteria for Adaptation

This is the background we need to understand the phylogenetic definition of adaptation. In technical terms, this definition has been rephrased as follows: "With respect to the phylogeny [cladogram] depicted [figure 2.2], the complete statement would be the derived trait M1 [of knuckle-walking] arose at time (t) in the stem lineage of [gorillas and chimpanzees] taxa C, D, and E via selection for the derived function F1 with respect to the primitive trait M0 [of brachiation] with primitive function [swinging under branches] F0 in [gibbons and siamangs] taxa A and B" (Coddington 1988, 5; examples in brackets added by authors). In plain English, knuckle walking (M1) arose at nine million years ago (t) in the stem lineage or common ancestor of African apes via selection for terrestrial locomotion (F1) with respect to the primitive trait of brachiation (M0) in lesser apes' taxa (A and B).

This definition of adaptation is useful because it provides a *standard for evaluating alternative adaptive scenarios* (Coddington 1988). Only adaptive scenarios that specify the phylogeny of the character as well as likely selection pressures can be considered potentially valid. We call this the *phyletic standard*. Specification of all of these features in one taxonomic group meets the minimum phyletic standard for an adaptive hypothesis. Specification of these features in two distantly related groups that show similar independently evolved functions strengthens the hypothesis. We call this the *convergence standard* for demonstrating adaptation.

This *phyletic approach* is ideal for understanding the evolution of shared derived character states of hominoids, as for example, certain cognitive abilities shared by all the great apes (Parker and Gibson 1977), and cooperative hunting and territorial defense shared by chimpanzees and humans (Stanford 1999). Unfortunately, however, the phyletic approach does not apply to uniquely derived character states of humans. For example, food storage and provisioning, warfare, castes, slavery, and domestication are not amenable to the phyletic approach because these traits are only found in humans and so are uniquely derived characters.

An alternative analytic strategy, the *correlation approach*, is useful for deducing the adaptive significance of character states unique to a related group, but relatively widely distributed across many distantly related taxa (Brooks and McLennan 1991; Harvey and Pagel 1991). This approach involves searching for a correlate of the focal character state exhibited in all the species displaying the character. Gibson (1986), for example, has used this approach to show that extractive foraging in many distantly related mammals correlates with neocortical progression. We call this the *correlation standard* for determining adaptation through examination of convergent cases (see table 2.2 for a summary of these two methods). See Parker and McKinney (1999) for discussion of these methods.

Table 2.2. Cladistic (Phyletic) versus Correlational Criteria for Adaptive Hypotheses

	Cladistic (Phyletic) Criteria	Correlational Criteria
For definition of an adaptation	Shared derived character state (phyletic standard)	Uniquely derived character state
For identification of convergent adaptations	Convergent function in one distantly related group (convergence standard)	Convergent function in many distantly related groups (correlation standard)
For identification of associated features	Generation of adaptive hypothesis based on features associated with convergent functions	Generation of adaptive hypothesis based on features correlated with convergent functions
Example	Intelligent tool use in great apes, convergent function in cebus monkeys: extractive foraging associated in both taxa	Extractive foraging in many distantly related taxa: increased neocortical index associated in all

Now we can return to the issue of distinguishing adaptations from exaptations. Gould (2002), citing Arnold (1994), argues that the primary basis for distinguishing between the two is direct knowledge of historical sequences or inference of these sequences from examination of phylogenetic branching points (nodes). The trick is to see whether the new characteristic appears before, after, or during the appearance of new selection pressures. If the trait appears first, it is an exaptation; if it appears later, it is an adaptation. If it appears at the same time, adaptation is probable but not certain. For example, if aimed throwing and tool use in the common ancestor of chimpanzees and humans preceded bipedal walking in early australopithecines, they were exaptations for bipedalism. Since loss of divergent big toes and shortening of the foot appeared after bipedal walking, they are adaptations for this form of locomotion.

As indicated in chapter 1, we use these concepts for analyzing the plausibility of the various hominin evolutionary scenarios discussed in this book. For example, phylogenetic concepts help us recognize cases in which scenario builders have "explained" characters shared with other great apes, and therefore not uniquely derived at all (chapter 7). We also judge scenarios by their compatibility with information about fossil hominins and their artifacts, related behaviors and artifacts of living hunter-gatherers, and those of our closest living relatives, chimpanzees and bonobos.

3

Who Were Our Ancestors?

> In each great region of the world the living mammals are most closely related to the extinct species of the same region. It is therefore probable that Africa was formerly inhabited by apes closely allied to the gorilla and chimpanzee; and as these two species are now man's nearest allies, it is somewhat more probable that our early progenitors lived on the African continent than elsewhere. But it is useless to speculate on this subject for two or three anthropomorphous apes . . . existed in Europe during the Miocene.
>
> —*Darwin 1871, 520*

This book is about scenarios of the evolution of uniquely human characteristics. Before we begin to discuss these scenarios, we consider recent views on the taxonomy of human ancestors and our living closest relatives, the African apes. Then we discuss recent ideas about the identity, anatomy, and behavior of our fossil ancestors. All the fossils are placed within the framework of geological time (Fleagle 1999); see table 3.1 for customary divisions into periods and epochs.

Since Darwin's formulation of descent with modification and genealogical relations among species (Darwin 1859), most taxonomists have based their classification of organisms on their understanding of evolutionary relationships among species, that is, on their family trees. Because, in Darwin's day, only one fossil human ancestor was recognized (Neanderthal), taxonomies of primates were limited to living species. With the discovery of fossil apes and hominins, primate taxonomies have been expanded to include extinct species.

Table 3.1. Geologic Timetable for the Cenozoic Era, Including Major Events in Primate and Hominin Evolution

Era	Period	Epoch	Age	Major Events in Primate and Hominin Evolution
Cenozoic	Neogene	Quaternary	0.01–present	Industrial revolution; agricultural revolution; development of writing; Neolithic culture; Early Stone Age
		Pleistocene	1.8–0.01 million years ago	Anatomically modern humans; Late Stone Age; Middle Stone Age
		Pliocene	5–1.8 million years ago	Hominins radiate
		Miocene	24–5 million years ago	First Old World monkeys; first apes; first hominins?
	Paleogene	Oligocene	38–24 million years ago	First Old World primates; first New World monkeys
		Eocene	54–38 million years ago	First prosimians
		Paleocene	65–54 million years ago	First primates

Source: Adapted from Fleagle, J. G. 1999. *Primate adaptation and evolution.* New York: Academic Press. Copyright © Academic Press 1999.

Until the 1960s, taxonomies had been based on comparative studies of the anatomical or behavioral characteristics of related species. Specifically, the presence of the same new characteristics in two or more similar species, which cladists call *shared derived characters*, were and are seen as evidence of descent from a common ancestor (Wiley 1981). For example, knuckle walking in gorillas, chimpanzees, and bonobos might be a shared derived character because it is common to them but not to orangutans or gibbons. Likewise, nonhuman African apes share other derived characters, including features of the skull, thin enamel on their molar teeth, and similar facial and gestural expressions of emotion. Of course, they also share such derived characters as nest building and tool use with orangutans, and such shared derived characters as the elongated canine teeth and sharpening premolar teeth with Old World monkeys and lesser apes.

PRIMATE TAXONOMY

As discussed in chapter 2, most systematists prefer taxonomies that are consistent with evolutionary relationships among species. (See discussion of shared derived versus ancestral shared characters as different forms of homology in chapter 2.) This phyletic or cladistic approach continues today to be the basis for classifying fossil apes and humans. Beginning in the 1950s, biologists began to use comparative analysis of protein molecules and then of DNA to construct their family trees of living species, and hence their taxonomies (Goodman 1963a, 1963b; Sarich 1968; Sarich and Cronin 1976). Data from this new approach were consistent with the older approach at the level of the superfamily (superfamily Hominoidea; see table 3.2). These taxonomies are indistinguishable above the family level. Beginning at the family level, however, these data imply a different taxonomy (Groves 1989).

In older taxonomies, the superfamily Hominoidea is divided into three families: the Hylobatidae (the lesser apes: gibbon and siamang), the Pongidae (the great apes: orangutan, gorilla, chimpanzee, and bonobo), and the Hominidae (all humans and their bipedal ancestors [traditional taxonomy in table 3.2; also see figure 3.1a]). This classification was based on anatomical similarities that suggest the great apes are more closely related to one another than any were to humans.

New Molecular-Based Taxonomy

In contrast, molecular data show that humans and the African apes (gorillas, bonobos, and chimpanzees) are more closely related to one another than any is to orangutans. More accurate evolutionary relationships among members of the Hominoidea based on the molecular approach imply changes in taxonomy: One change is that there are only two families (the Hylobatidae [the lesser apes] and the Hominidae [humans *and* the great apes]; recent taxonomy is in table 3.2) instead of three families within the Hominoidea. By subsuming all the great apes and humans under the Hominidae, this family becomes a valid (monophyletic) clade; it includes all of the descendants of the common ancestor of the great apes and humans (see table 3.2 and figure 3.1b). In this revised taxonomy, within the family Hominidae, there are two subfamilies: the Ponginae (orangutans) and the Homininae (the African great apes and humans).

This classification is based on the current consensus, anchored in molecular data, that the African apes (gorillas, chimpanzees, and bonobos) and humans share a recent common ancestor and are more closely related to each other than any is to the orangutan. Therefore, humans and the African apes are now classified together as members of the subfamily Homininae (figure

Table 3.2. Alternative Taxonomies for Extant Species of the Superfamily Hominoidea

	Traditional Taxonomy			Recent Taxonomy (after Groves 1989)			
Kingdom	Animalia			Animalia			
Phylum	Cordata			Cordata			
Class	Mammalia			Mammalia			
Order	Primates			Primates			
Suborder	Anthropoidea			Anthropoidea			
Infraorder	Catarrhini			Catarrhini			
Superfamily	Hominoidea			Hominoidea			
Family	Hylobatidae	Pongidae		Hylobatidae	Hominidae		
Subfamily			Homininae		Ponginae	Homininae	
Tribe						Gorillini	Panini	Hominini
Genus	Hylobates	Pongo Gorilla Pan	Homo	Hylobates	Pongo	Gorilla	Pan	Homo

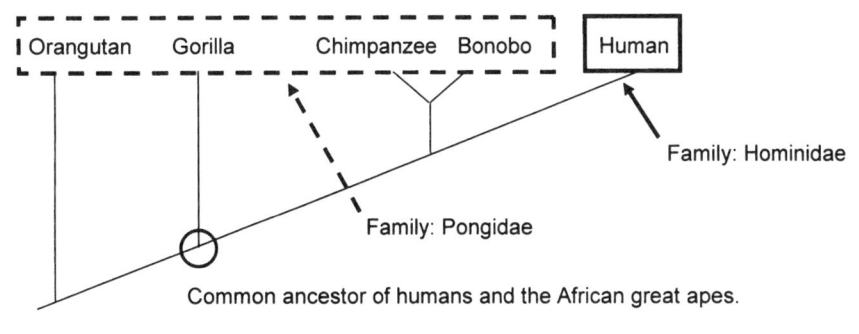

FIGURE 3.1A.
Cladogram showing the taxonomic relationships in the traditional taxonomy.

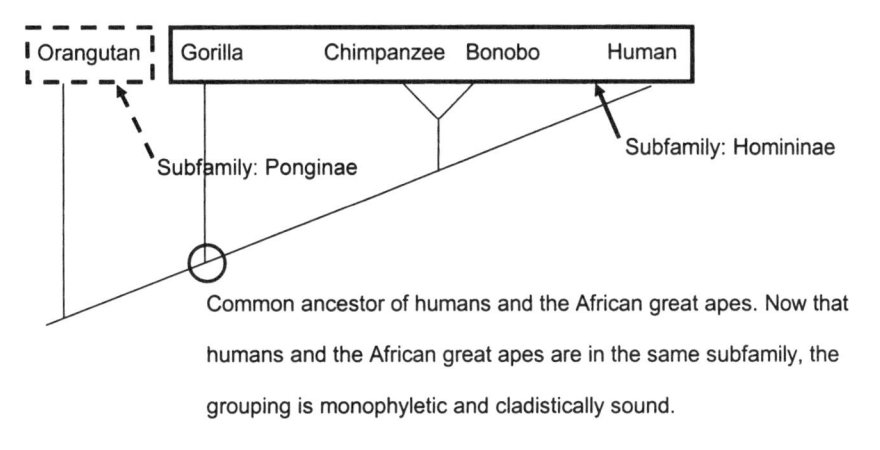

Labels in figure: Orangutan, Gorilla, Chimpanzee, Bonobo, Human; Family: Hominidae; Family: Pongidae; Common ancestor of humans and the African great apes. Because humans are placed in their own family, this grouping (in dashed box) is paraphyletic.

FIGURE 3.1B.
Cladogram showing the taxonomic relationships in the recent taxonomy.

Labels in figure: Orangutan, Gorilla, Chimpanzee, Bonobo, Human; Subfamily: Ponginae; Subfamily: Homininae; Common ancestor of humans and the African great apes. Now that humans and the African great apes are in the same subfamily, the grouping is monophyletic and cladistically sound.

3.1b). (Orangutans are slightly more distant relatives of all the Homininae, and therefore remain in the older subfamily Ponginae.)

Humans and their extinct bipedal relatives (e.g., the australopithecines, archaic forms of *Homo*) form their own clade within the Homininae (they share a common ancestor to the exclusion of the African apes). Therefore, the revised taxonomy necessitates another level that distinguishes among the members of the subfamily Homininae. This problem has been addressed by dividing the Homininae into three tribes (Gorillini, Panini, and Hominini).

In the revised taxonomy, the *tribe* Hominini has taken the place of the family Hominidae in referring to the group of habitually bipedal primates that includes modern humans, their extinct bipedal ancestors, and related groups that have lived after our lineage split from the lineage leading to the African apes. Colloquially, this group now is referred to as hominins rather than hominids. (We use "hominins" throughout the book except when describing scenarios by earlier authors who use the older term, "hominids.") Hominini is still a cladistically valid grouping since all hominins share a common ancestor after the African apes branched off (Groves 1989).

Within this consensus, however, taxonomists continue to argue over how many genera and species of hominini there are. Taxonomists who favor larger groupings with fewer taxa following Ernst Mayr (Mayr 1950) are known as lumpers, and those who favor smaller groupings with more taxa (Tattersall 1986) are known as splitters. Advocates of these two positions base their judgments on different definitions of species and hence different criteria for classifying them.

STAGES OF HUMAN EVOLUTION

To avoid these controversial details, we use the following nontechnical terms for our ancestors throughout the book except when details are warranted. Each category (except the LCA) represents a new evolutionary stage or grade of hominin.

1. *Last common ancestor (LCA)* is the term we use for our last common ancestor with the African apes, most likely a great ape of the genus *Dryopithecus*. Others have suggested the term stem ape (Tanner 1987) or *Pan prior* (Wrangham 2001). They all lived in Africa, but none were habitual bipeds.
2. *Basal hominins* is the term we use to refer to the earliest fossils in our tribe, dating from about seven to four million years ago, that is, from zero to two million years after our last common ancestor with African apes. These include *Sahelanthropus*, *Orrorin tugenensis*, and *Ardipithecus ramidus*. They all lived in Africa. The skeletal material associated with the species listed as "basal hominins" *suggests* that these species may have been upright (e.g.,

forwardly placed foramen magnum), but conclusive skeletal evidence of bipedal locomotion in these species is currently lacking.
3. *Early human ancestors* refers to fossils in our tribe dating from about 3.5 to about 2.4 million years ago. These include the older subgroup of *Kenyanthropus platyops, Australopithecus anamensis, A. bahrelghaszli,* and *A. afarensis*, and the younger subgroup including *A. africanus, A. garhi, Paraustralopithecus aethiopicus, Paranthropus* species, and *Homo rudolfensis* and *H. habilis*. Some of these species made and used simple cobblestone tools. They all lived in Africa. Species designated as "early human ancestors" are differentiated from "basal-humans" both by their lesser ages and by displaying skeletal morphology necessary for habitual bipedal locomotion.
4. *Middle-period human ancestors* refers to fossils in our tribe dating from about 1.9 million years ago to about three hundred thousand years ago; the best known of these are *H. erectus* and *H. ergaster*. They made and used biface stone tools. Some groups of these ancestors left Africa and moved into Georgia, Java, and China. We separate these species from "early human ancestors" because they are larger, more efficient bipeds whose brain size is intermediate between those of African apes and modern humans.
5. *Recent human ancestors* refers to earlier archaic humans including *H. antecessor* and *H. neanderthalensis*, that arose between 800,000 to 250,000 years ago, as well as fossils in our species, anatomically modern *Homo sapiens*, which arose about 160,000 years ago. Some of these ancestors left Africa and moved into the Middle East, Southeast Asia, and Europe. We separate these species from "middle-period human ancestors" because they have brain sizes and cultural remains similar to those of preliterate living humans.

Finally, *human ancestors* is a general term we use for all of the known fossils dated between the split between ancestral African apes and the earliest hominins and the appearance of anatomically modern humans (that is, all members of the tribe Hominini). Current evidence suggests that basal humans and early human ancestors lived exclusively in Africa. It also suggests that middle-period human ancestors, that is, *Homo ergaster* (previously referred to as *Homo erectus*), were the first to leave Africa, migrating from there into Java and China (Curtis, Swisher, et al. 2000).

All the fossil species that lived after the first basal hominin split from the common ancestor of African apes are Hominini (table 3.3), but only one species among those living at a given time actually gave rise to the lineage leading to modern humans. In most cases, anthropologists do not know which species this was, and may never know. This is because whereas every living species had an ancestor, not all fossil species had descendants. In fact, most of them became extinct without issue.

Who Was Our Last Common Ancestor?

Until recently these arguments remained unresolved because anthropologists have found few fossil apes, and of those even fewer with limb bones. In recent years, both anatomical evidence from new fossils and DNA evidence from living primates has shown conclusively that the last common ancestor was an African ape that lived about six to eight million years ago (Goodman 1963a, 1963b; Sarich 1968; Sarich and Cronin 1976; Cronin, Sarich, et al. 1984; Ruvolo 1994). Recently some new fossil apes from this time period have helped clarify the family tree of apes (Begun 1992; Begun, Ward, et al. 1997; Kordos and Begun 2002). See figure 3.2 for a family tree of apes.

Surprisingly, given their current geographical distribution in Africa, the last common ape ancestor of African apes and humans apparently lived in Europe or Eurasia about six to nine million years ago during the late Miocene, and not in Africa. Lazlos Kordos and David Begun (2002) describe Miocene ape fossils of the genus *Dryopithecus* from Rudabánya, Hungary, which they classify as a relative of the ancestors of African apes and humans on the basis of many shared derived characters in the face and dentition. They describe the limb bones of this creature as being similar to those of living apes, indicating suspensory behavior, that is, hanging and moving under branches in a forest setting.

Given the absence of ape fossils in Africa before about four million years ago in the late Miocene, Kordos and Begun (2002) suggest that African apes originated in Eurasia and migrated into Africa when climate changes in Europe forced them south. The authors suggest that a relative of this group dispersed from western Europe or the eastern Mediterranean into Africa through western Asia sometime between six and nine million years ago and radiated into the African apes and human ancestors. Others have suggested a similar scenario based on molecular data (Stewart and Disotell 1998).

Table 3.3. Fossil Hominin Species

Name	Authors	Geologic Date	Country	Type Specimen
Sahelanthropus tchadensis	Brunet, Guy, et al. (2002)	7 million years ago	Chad, East Africa	TM 266-01-060-1
Orrorin tugenensis	Senut, Pickford, et al. (2001)	6 million years ago	Ethiopia, East Africa	BAR 1000'00
Ardipithecus kadabba	Haile-Selassie (2001); Haile-Selassie, Suwa, et al. (2004)	5.8–5.2 million years ago	Ethiopia, East Africa	ALA-VP 2/10
Ardipithecus ramidus	T. White, Suwa, et al. (1994)	4.4 million years ago	Ethiopia, East Africa	ARA-VP 6/1
Australopithecus anamensis	M. Leakey, Feibel, et al. (1995)	3.9–4.2 million years ago	Kenya, East Africa	KNM-KP 29281
Kenyanthropus platyops	M. Leakey, Spoor, et al. (2001)	3.5–3.2 million years ago	Kenya	KNM-WT 40000
Australopithecus bahrelghazali	Brunet, Beauvilain, et al. (1996)	3.5–3.0 million years ago	Chad, East Africa	KT12/H1
Australopithecus afarensis	Johanson, White, et al. (1978)	3.2 million years ago	Ethiopia, East Africa	LH 4
Australopithecus garhi	Asfaw, White, et al. (1999)	2.5 million years ago	Ethiopia, East Africa	ARA-VP-12/130
Australopithecus africanus	Dart (1925)	2.5 million years ago	Taungs, South Africa	Taungs Baby
Paraustralopithecus aethiopicus	Arambourg and Coppens (1968)	2.5 million years ago	Kenya, East Africa	Black Skull (WT17000)
Paranthropus robustus	Broom (1938)	2.0 million years ago	Swartkranz, South Africa	SK 48
Paranthropus boisei	L. Leakey (1959)	1.76 million years ago	Olduvai, Tanzania, East Africa	Dear Boy OH 5
Homo habilis	L. Leakey, Tobias, et al. (1964)	2.5 million years ago	Olduvai, Tanzania, East Africa	OH 7 & 13
Homo rudolfensis	Alexeev (1986)	2.9 million years ago	Kenya, East Africa	KNM-ER 1470
Homo erectus	DuBois (1892)	1.6 million years ago	Java, Asia	Trinil 2
Homo ergaster	Groves and Mazak (1975)	1.6 million years ago	Kenya, East Africa	KNM-ER 992
Homo antecessor	Bermudez de Castro, Arsuaga, et al. (1997)	800,000 years ago	Spain, Europe	ATD6-5
Homo neaderthalensis	King (1864)	125,000 years ago	Germany, Europe	Neandertal 1
Homo sapiens	Linnaeus (1758)	60,000 years ago	Global	

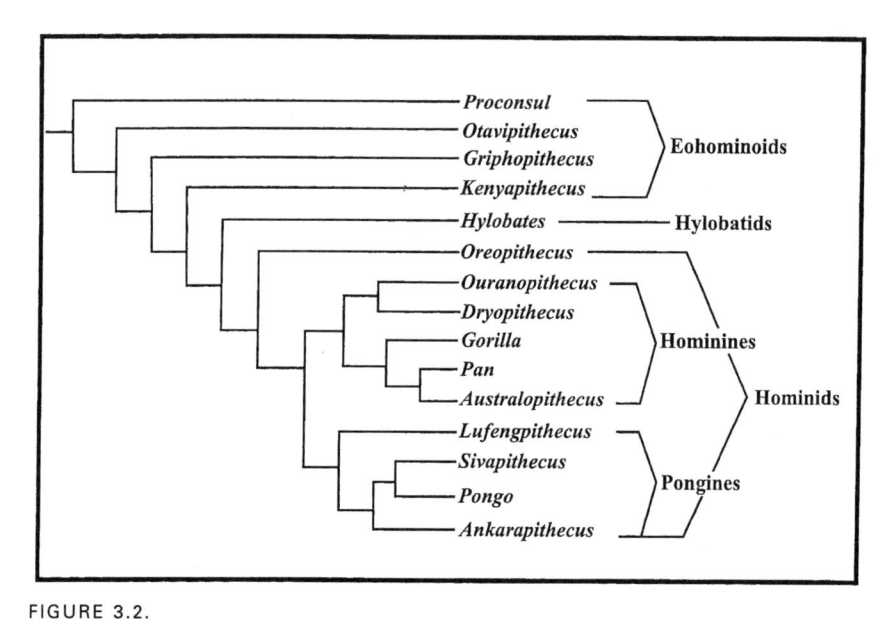

FIGURE 3.2.
Family tree of the fossil apes.
Source: Kordos, L., and D. R. Begun. 2002. A late Miocene subtropical swamp deposit with evidence of the origin of African apes and humans. *Evolutionary Anthropology* 11:45–57. Copyright © 2002 John Wiley & Sons, Inc. Reprinted with the permission of Wiley-Liss, Inc., a subsidiary of John Wiley & Sons, Inc.

Basal Hominins

In recent years, paleoanthropologists have discovered and named several late Miocene primates that they consider to be hominins: *Sahelanthropus*, *Orrorin*, and *Ardipithecus*. These primates display traits that may indicate bipedality. In 2002, an international team led by Mark Brunet discovered the skull of a six- to seven-million-year-old potential hominin, *Sahelanthropus tchadensis* (Brunet, Guy, et al. 2002), in Djurab Desert, Chad. This find is remarkable for two reasons. First, if *S. tchadensis* is determined to be bipedal, the discovery will push back the date for the earliest hominin very close to when molecular studies place the last common ancestor of chimpanzees and hominins (Brunet, Guy, et al. 2002).

Second, the discovery of a probable hominin in central Africa suggests that these creatures were more widely distributed than previously assumed (Brunet, Beauvilain, et al. 1996). Although no postcranial remains were recovered, several aspects of the base of the skull, including the length, horizontal orientation, and anterior placement of the foramen magnum, suggest bipedality.

Before the 2002 discovery of *Sahelanthropus tchadensis*, the earliest potential hominin was *Orrorin tugenensis* (Pickford and Senut 2001; Senut, Pickford, et al. 2001), discovered at Lukeino Formation in Kenya in 2000, dated approximately six million years ago. Unlike the *Sahelanthropus tchadensis* discovery, *Orrorin tugenensis* consisted of some postcranial material. Aspects of the femur, including the position and relative size of the femoral head and the length of the femoral neck, suggest bipedalism (Pickford and Senut 2001; Senut, Pickford, et al. 2001). However, other aspects of the postcrania, including long, curved phalanges, suggest that *Orrorin tugenensis* also relied heavily on arboreal locomotion.

In 1994, Tim White and colleagues discovered what was then considered the earliest possible hominin, dated to 4.4 million years ago. Initially named *Australopithecus ramidus* (T. White, Suwa, et al. 1994), this specimen was elevated later to its own genus, *Ardipithecus ramidus*, because it appears to be considerably more primitive than other members of the genus *Australopithecus*. Later discoveries of fossils attributed to *Ardipithecus ramidus* may represent a different subspecies (Haile-Selassie 2001). Currently, *Ardipithecus ramidus* is only known from sites in the Middle Awash of Ethiopia. As with *Sahelanthropus tchadensis*, the primary morphological trait suggestive of bipedalism is a shortened base of the skull and a forwardly placed foramen magnum (T. White, Suwa, et al. 1994). Postcranial bones attributed to *Ardipithecus ramidus* have yet to be described.

In addition to morphology related to bipedalism, these species display a variety of other characteristics later found in australopithecines, including thick enamel on their molar teeth and reduced canine and premolar teeth. Unfortunately, because few of the same bones and teeth are found among these fragmentary fossils, comparisons are difficult (Begun 2004). Interpretations of their behavior and life history have been proposed in various scenarios primarily on the basis of their similarities to apes.

Early Human Ancestors

There is an extensive fossil record of early human ancestors in East and South Africa between 3.9 and 1.75 million years ago. These include *Kenyanthropus* (M. Leakey, Spoor, et al. 2001) and the early australopithecines,

Australopithecus anamensis (M. Leakey, Feibel, et al. 1995), *A. afarensis,* and *A. bahrelghazali* (Brunet, Beauvilain, et al. 1996) living between 3.9 and 2.4 million years ago. It also includes *A. garhi* (Asfaw, White, et al. 1999); the small-toothed "gracile" later australopithecines, *A. africanus* (Dart 1925), the large-toothed "robust" *Paranthropus robustus* (Broom 1938), and *P. boisei* (formerly *Zinjanthropus* and *Australopithecus*) (L. Leakey 1959); and hyper-robust *A. aeithiopicus,* which was initially in the genus *Paraustralopithecus* (Arambourg and Coppens 1968) living between 2.4 and 1.75 million years ago. These species show enlarged molars, reduced size of canine teeth compared to chimpanzees, and various forms of primitive bipedalism suggesting that at least some of them still climbed trees (Stern and Susman 1983; Susman, Stern, et al. 1985). Although they showed considerable sexual dimorphism (table 3.4), these were all small creatures (less than fifty kilograms) whose brain sizes only slightly exceeded those of African apes (less than 550 cubic centimeters; McHenry 1988).

Of the australopithecines, only *A. garhi* is definitively associated with worked stone tools (Asfaw, White, et al. 1999; Heinzelin, Clark, et al. 1999). The size and shape of its thumb suggest that *Paranthropus robustus* from South Africa (Susman 1998) had the capacity for stone tool production.

In contrast, *Homo habilis* (L. Leakey, Tobias, et al. 1964) and *Homo rudolfensis* (originally *Pithecanthropus rudolfensis*; Alexeev 1986), living in Africa between 2.9 and 2.4 million years ago, have larger brains than australopithecines, and are associated with simple worked stone tools (L. Leakey, Tobias, et al. 1964; R. Leakey 1973). *Homo habilis* has been controversial from its designation (Wood 1992). Paleoanthropologists were surprised to discover a tiny

Table 3.4. Estimated Body Weights and Heights for Various Hominins

	Body Weight (kg)		Stature (cm)	
	Male	Female	Male	Female
A. afarensis	45	29	151	105
A. africanus	40	30	138	115
P. robustus	40	32	132	110
P. boisei	49	34	137	124
H. habilis	52	32	157	125
H. ergaster	63	42	180	160
H. sapiens	65	54	175	161

Source: Adapted from Klein, R. G. 1999. *The human career: Human biological and cultural origins.* Chicago: University of Chicago Press.

skeleton of this species with arm and leg proportions similar to Lucy (*A. afarensis*) (Johanson 1986). Some investigators have concluded that *H. habilis* includes more than one species. Some of the fossils in its hypodym have been designated *H. rudolfensis*. Bernard Wood (Wood 1992) has suggested that the designation genus *Homo* should be reserved for those fossils with enlarged brains and fully bipedal walking. He has questioned whether *H. habilis* was ancestral to *H. erectus* because they were contemporaneous. Fred Spoor and colleagues (Spoor, Leakey, et al. 2007) recently reported the discovery of an *H. habilis*, dated 1.44 million years ago, living in association with *H. erectus*. They also suggest that the new, younger age of the *H. habilis* and the coexistence of the two species cast doubt on the idea that *H. habilis* was ancestral to *H. erectus*.

Middle-Period Human Ancestors

Middle-period human ancestors include *Homo erectus* (Weidenreich 1939) and *Homo ergaster* (Groves and Mazak 1975) living from 1.9 million years ago to about three hundred thousand years ago in Africa, Java, and China. The first *H. erectus* fossil was found on the island of Java, at the site of Trinil, by Eugene DuBois in 1891 (DuBois 1892). Since that time, fossils referred to as *H. erectus* have been uncovered in China (e.g., Zhoukoudian cave near Beijing [Weidenreich 1939], Africa (e.g., Olduvai Gorge, Tanzania [Day 1971; Rightmire 1979]), Swartkrans, South Africa (Clarke, Howell, et al. 1970), and Lake Turkana, Kenya (F. Brown, Harris, et al. 1985; J. Brown and Pimm 1985; R. Leakey and Walker 1985a; R. Leakey and Walker 1985b); and more recently in west Asia (e.g., Dmanisi, in Georgia [Gabunia and Vekua 1995]), and Ceprano, in Italy (Ascenzi, Bidittu, et al. 1996). Although earlier finds suggest that *H. erectus* was less sexually dimorphic than australopithecines, recent discovery of a small female skull suggests that this species was highly dimorphic (Spoor, Leakey, et al. 2007).

Recently, however, the classification of *H. erectus* has become more complicated. Owing to the high degree of variation in this widespread species, many paleoanthropologists think the fossils labeled *H. erectus* actually represent two species (Tattersall 1986; Groves 1989; Wood 1991). Others disagree (Rightmire 1990; Brauer and Mbua 1992; Kramer 1993). Under this new classification, the name *Homo erectus* is reserved for the Asian specimens (i.e., those in China and Java), while specimens in Africa, Europe, and west Asia have been given the species name *Homo ergaster* (Groves and Mazak 1975).

Whereas the two species are similar in many morphological respects, they differ, for example, in that *H. ergaster* has a higher cranial vault with thinner cranial vault bones, no sagital keel, and a smaller cranial capacity when compared to the more narrowly defined *H. erectus*.

Homo ergaster and its apparent descendant, *H. erectus*, display several significant morphological, behavioral, and technological changes compared to earlier hominins. The 1984 discovery near Lake Turkana of the Nariokotome boy (aka Turkana Boy) revealed differences in morphology between *H. ergaster/erectus* and earlier hominins (including *H. habilis*). The individual was immature, yet his height was five feet at death. It is estimated that he would have topped six feet and weighed 140 pounds had he lived to sexual maturity (F. Brown, Harris, et al. 1985; R. Leakey and Walker 1985a). These height and weight estimates are within the range of modern human variation, and are much larger than those of earlier hominins (Klein 1999; table 3.4). In addition to more modern height and weight, the Nariokotome boy also displayed modern limb proportions: long legs in comparison to arm length (F. Brown, Harris, et al. 1985; R. Leakey and Walker 1985a). This is a marked shift from earlier hominins, who retain the arboreal adaptation of relatively long arms in relation to leg length, for example, *A. afarensis* (Jungers 1982).

Perhaps these morphological changes enabled *H. ergaster* to colonize parts of the Old World outside of Africa. If the dates are correct (and there is controversy about them), *H. ergaster* reached Java by 1.9 million years ago (Swisher, Curtis, et al. 1994; Curtis, Swisher, et al. 2000) and Dmanisi (in the Republic of Georgia) 1.6 to 1.8 million years ago, perhaps only two hundred thousand to four hundred thousand years after their initial appearance in Africa (Gabunia and Vekua 1995). The shift in *H. ergaster*'s morphology toward more efficient bipedalism and less reliance on trees may have enabled its migration. In any event, there is no good evidence for hominins outside of Africa before the appearance of *H. ergaster*.

Although earlier hominins (e.g., *H. habilis* and *A. garhi*) made and used simple stone tools, the appearance of *H. ergaster* marks a shift to a new tool assemblage, the Acheulean. The bifacially flaked Acheulean tools represent significant innovation over earlier tool kits, that is, the Oldowan (Schick and Toth 1994). Another technological advance associated with *H. ergaster/erectus* is the first evidence of controlled use of fire. *Homo ergaster/erectus* apparently ate more animal protein than earlier hominins, but whether they obtained it by scavenging or hunting is a source of debate (Klein 1987; also see chapter 4).

Recent Human Ancestors

Recent human ancestors include fossils ascribed to *Homo sapiens* and earlier archaic forms of *Homo* possibly dating from as early as eight hundred thousand years ago, which are clearly distinct from *H. ergaster/erectus*. Paleoanthropologists disagree about the number of archaic *Homo* species living during this period (McBrearty and Brooks 2000). Because of this, hominin specimens clearly belonging to neither *Homo ergaster/erectus* nor *Homo sapiens* are often grouped together and referred to as "archaic *Homo sapiens*." For clarity, some distinguish between older archaic *H. sapiens* and younger archaic *H. sapiens*. Some of these fossils have been classified as separate species.

Homo antecessor, dated to approximately eight hundred thousand years ago, was discovered in 1995 and 1996 at the Gran Dolina cave site of Atapuerca Hills, Spain (Bermudez de Castro, Arsuaga, et al. 1997; Carbonell, Esteban, et al. 1999). Initially, their discoverers considered them to represent a primitive form of *Homo heidelbergensis* (Carbonell, Bermudez de Castro, et al. 1995). Upon discovery of more facial skeletal material displaying a unique combination of derived and primitive *Homo* traits, however, the researchers altered their assessment of the fossil material and elevated it to its own species, *Homo antecessor* (Bermudez de Castro, Arsuaga, et al. 1997).

The species designation *Homo heidelbergensis* was proposed by Otto Schoetensack (1908) after he discovered a nearly complete mandible (the Mauer jaw) near the village of Mauer, just south of Heidelberg, Germany, in 1908. This type specimen for *Homo heidelbergensis* has been dated to about five hundred thousand years ago. Like the *Homo antecessor* finds discussed above, the Mauer jaw displays a variety of primitive and derived features (Stringer, Hublin, et al. 1984). Most other archaic *Homo* specimens from Europe (e.g., Vértesszöllös, Boxgrove) have also been placed in *Homo heidelbergensis*. *Homo heidelbergensis* may be ancestral to *H. neanderthalensis*.

Homo neanderthalensis is a more homogenous group of younger archaic *Homo* known from European and west Asian sites dated between about one hundred and thirty thousand and thirty-five thousand years ago. The first widely recognized Neanderthal remains were found in Feldhofer Cave in the Neander Valley, Germany, in 1856 (King 1864). Although earlier discoveries of Neanderthal fossils had been made (a child's skull discovered at Engis Cave in Belgium in 1830 and a woman's skull discovered at Forbes's Quarry in Gibraltar (Spencer 1984; Stringer, Hublin, et al. 1984), these went largely unrecognized because the fossils' archaic morphology was not immediately clear. In

1864, the Irish anatomist William King determined that the Neanderthal fossils were sufficiently different from modern humans to warrant their own species designation: *Homo neanderthalensis* (King 1864).

Morphological differences between Neanderthals and anatomically modern humans include the large face, low-slung braincase, large teeth, and heavy postcranial skeleton. As a group, Neanderthals display a number of *autapomorphies*, unique traits that differentiate them from other hominins (Stringer, Hublin, et al. 1984; Holloway 1985; Franciscus and Trinkaus 1995). It is significant that many of these traits show up in early infancy (Minugh-Purvis, Radovcic, et al. 2000)

In addition to displaying a unique morphology, Neanderthals are associated with the distinctive Mousterian tool technology, named for the Le Moustier rock shelters in France (Stringer and Gamble 1993). Other Middle Stone Age industrial complexes made during this time period outside the geographic range of the Neanderthals (e.g., North Africa), have been referred to as Mousterian (Van Peer 1991) or given their own names, for example, Aterian (Ferring 1975). There are a few Neanderthal fossils associated with non-Mousterian industrial complexes so that the strict boundary of Neanderthal/non-Neanderthal tool technologies is not as clear as it might seem.

There has been considerable debate regarding the fate of the Neanderthals. The majority of paleoanthropologists think anatomically modern humans replaced Neanderthals, with little or no interbreeding (Stringer and Gamble 1993). A few anthropologists believe Neanderthals were the direct, local ancestors of later anatomically modern human populations (Hrdlicka 1927; Wolpoff, Hawks, et al. 2001).

This idea of Neanderthal extinction is based on at least several lines of evidence: first, anatomical differences between Neanderthal and modern *H. sapiens*; second, the long coexistence of Neanderthals and modern *H. sapiens* in the Middle East; third, the sudden disappearance of Neanderthals about thirty thousand years ago; fourth, anatomical differences that appear in infant Neanderthals suggest a genetic rather than environmental cause (Minugh-Purvis, Radovcic, et al. 2000); fifth, archaeological differences in tool cultures, suggesting that Neanderthals coexisted with anatomically modern humans for thousands of years in the Middle East and then disappeared suddenly (Stringer and Gamble 1993; Rak, Kimbel, et al. 1994; Tattersall 1995; Krings, Stone, et al. 1997; McBrearty and Brooks 2000; Minugh-Purvis, Radovcic, et al. 2000; Lewis-Williams 2002). Finally, genetic differences between the DNA of Neanderthals and modern humans suggest that Neanderthals diverged long before anatomically modern humans evolved (Krings, Stone, et al. 1997). See table 3.3 for a list of human ancestors and their time spans.

Three lines of molecular research show that neither Neanderthals nor other archaic *Homo* species could have contributed DNA to the modern human gene pool (Krings, Stone, et al. 1997): first, a firm estimate of two hundred thousand years ago for the last common ancestor for human mitochondrial DNA (Cann, Stoneking, et al. 1987; Vigilant, Stoneking, et al. 1991); second, a firm estimate of sixty thousand years ago for the last common African ancestor for human Y chromosomes (Underhill, Shen, et al. 2000; Wells 2002); third, a firm estimate of the age of Neanderthal mitochondrial DNA having a last common ancestor with modern humans five hundred thousand years ago (Krings, Stone, et al. 1997). Moreover, fossil evidence for the earliest anatomically modern humans has been found only in Africa (most recently in Ethiopia) and dated to one hundred and sixty thousand years ago (T. White, Asfaw, et al. 2003). The hominins who replaced the Neanderthals in Europe and western Asia and other archaic *Homo* populations in other parts of the Old World, were members of our own species, anatomically modern *Homo sapiens*.

These lines of evidence settle centuries-long arguments over the fate of the Neanderthals. They also settle the decades-long arguments over the fates of other archaic human populations throughout the Old World. Specifically, they support a modified version of the Out of Africa (Stringer 1984; Stringer 1989; Stringer and Gamble 1993; Stringer 1996) over the Regional Continuity (or Multiregional Continuity) (Coon 1963; Wolpoff, Thorne, et al. 1984; Wolpoff, Zhi, et al. 1984). The Out of Africa model says that anatomically modern *Homo sapiens* evolved from archaic *Homo* populations in Africa and then dispersed from their ancestral home, replacing archaic populations they found in other parts of the Old World. The Regional Continuity model, on the other hand, hypothesizes that anatomically modern humans evolved from local archaic *Homo* populations in many locations throughout the Old World.

CONCLUSIONS

In the nearly century and a half since Darwin wrote *The Descent of Man* (Darwin 1871), paleoanthropologists have unearthed thousands of bones of human ancestors around the Old World and named scores of hominin species.

These include candidates for the last common ancestor and basal humans, as well as early human ancestors, middle-period human ancestors, and recent human ancestors. The chronology and biogeography of these species, combined with molecular data from recent populations, suggest a pattern of basal hominin origins and adaptive radiation in the Miocene Epoch in Africa, followed by at least four other adaptive radiations and two or three migrations out of Africa:

1. the first adaptive radiation, of early human ancestors (australopithecines and early *Homo*), occurring in Africa in the Pliocene Epoch between three and two million years ago;
2. followed by the African origin of middle-period human ancestors, *Homo ergaster*, and the first migration and adaptive radiation in Asia in the Pleistocene Epoch between two million and three hundred thousand years ago;
3. followed by another adaptive radiation of recent human ancestors, archaic *Homo* species in Africa, Europe, and Asia;
4. and finally, the origin of anatomically modern *H. sapiens* in Africa about one hundred and sixty thousand years ago, followed by migration throughout the Old World beginning ninety thousand to sixty thousand years ago, leading to the extinction of earlier remnant populations.

Anthropologists have reconstructed the locomotor patterns, diets, and body and brain sizes of these species, as well as their habitats and the technologies they used. In the following chapters, we discuss various scenarios anthropologists have proposed for the evolution of uniquely human characteristics. Whenever possible, we discuss new (derived) characteristics associated with each stage or grade of hominin evolution leading to modern humans. Our general strategy is to follow Darwin's lead in looking for evolutionary continuities between ancestor-descendant lineage by using comparative data on modern human tribal groups and our closest living relatives, African apes.

II
ORIGINS MODELS

4

Origins of Human Subsistence

> He [man] has invented and is able to use various weapons, tools, traps &c. with which he defends himself, kills or catches prey, and otherwise obtains food. He has made rafts or canoes for fishing or crossing over to neighbouring fertile islands. He has discovered the art of making fire, by which hard and stringy roots can be rendered digestible, and poisonous roots or herbs innocuous. The discovery of fire, probably the greatest ever made by man, excepting language, dates from before the dawn of history.
>
> —*Darwin 1871, 432*

Traditionally, the first questions anthropologists ask about early hominins are, what did they eat and how did they acquire their food? Did they eat meat or tubers or termites or seeds? If they ate meat, which animals did they eat, and did they hunt or scavenge them? Because diet is the keystone of survival and of a species' role in its ecological communities, it makes sense that it is a key feature in many scenarios of hominin origins. Likewise, it makes sense that most of these scenarios focus on natural selection, though a few mention sexual selection. As background for our discussion of subsistence scenarios, we need to identify the diet of the last common ancestor of chimpanzees, bonobos, and humans.

Bartholomew and Birdsell's Ecological Scenario for Hominid* Subsistence

Bartholomew and Birdsell (1953) propose to "extrapolate downward" from mammalian ecological data and to "extrapolate upward" from hunting and gathering peoples to the biological attributes of protohominids, australopithecines, and preagricultural humans. After summarizing such basic ecological concepts as population equilibrium and trophic levels, they note that an important corollary of these concepts is that "nutrition plays a primary role in determining the major functional adaptations of animals" (486). They discuss the following topics in relation to abstract protohominids and australopithecines: dentition, tools, diet, sociality, territoriality, and population equilibrium.

Regarding protohominids, they begin with a discussion of terrestrial bipedalism, noting that it is a critical feature of a "previously unexploited mode of life" (Bartholomew and Birdsell 1953, 482). They argue that its advantage was "the use of hands for efficient manipulation of adventitious tools such as rocks, sticks, or bones" (482). After briefly noting the presence of tool use in other species, they say that habitual upright posture made tool use obligatory for protohominids. As a consequence these were large enough (fifty to a hundred pounds) to avoid being prey except for large cats and pack hunting dogs and to exploit the "entire range of food size used by all other terrestrial mammals" (483) without specialization. They suggest that these creatures formed "relatively stable family groups" composed of "semi-permanent biological famil(ies), including offspring,"

---

*Hominids are now known as hominins

SUBSISTENCE STRATEGIES THROUGH STAGES OF HUMAN EVOLUTION
What Did the Last Common Ancestor Eat? Did They Hunt?

Our closest living relatives, chimpanzees, prefer fruits, but are omnivorous and very broad in their diets (Teleki 1974; Goodall 1986; Boesch and Boesch-Acherman 2000), and bonobos are intermediate between chimpanzees and gorillas, eating both herbs and fruits (Malensky, Kuroda, et al. 1994). In fact, Goodall (1986) notes that Wrangham identified 174 species of vegetable foods eaten by chimpanzees. Fruits, however, were the single largest component of the plant diet. Other plant parts and invertebrates such as ants and termites were "fall-back foods" for consumption in seasons when fruits were scarce.

and, like apes, had a long period of growth and maturation required for their "unique dependence on learning" (483). They also suggest that they had well-developed territoriality.

Bartholomew and Birdsell (1953) then discuss australopithecines as possible embodiments of protohominids, citing uncertainties about their dating and relationship to other hominids. They note that, like their protohominid model, these creatures were bipedal and probably used tools, based on tool use by great apes. In addition, they note the unique characteristic of reduced canines and incisors with nonsharpening premolars. They argue that the absence of stabbing teeth implies the killing and butchery of game by weapons and in intrasexual combat. Based on the body size of these creatures, the authors argue that crude tools such as clubs and digging sticks would have allowed these creatures to hunt virtually all small terrestrial mammals, aquatic reptiles, fish, nesting birds, eggs, and insects; to scavenge large kills of carnivores; and to excavate roots, tubers, and fungi, as well as collect fruits and nuts and buds and shoots. Citing sexual dimorphism in australopithecines, they say that they must have displayed aggressive male behavior and a sexual division of labor. Finally, the authors note that the australopithecines must have been subject to rapid replacement by highly evolved humans who were able to expand their range beyond the tropics and semitropics into temperate regions of the Old World. They end with a plea to anthropologists to stop overlooking "man's nutritive dependence upon the environment," and begin "long inhibited quantitative investigation of the relationship between man's population density and environmental factors" (496).

Although chimpanzees rely extensively on vegetation and invertebrates to fulfill their nutrient and caloric needs, studies of wild chimpanzees in Ivory Coast, Liberia, Senegal, Guinea, Uganda, Tanzania, and Zaire have revealed that chimpanzees are adept hunters (Boesch 1994). Detailed studies of hunting behavior in the semi-open woodlands of the Gombe Reserve in Tanzania (Teleki 1973; Stanford 1998) and in the Taï tropical rainforest in Ivory Coast (Boesch and Boesch 1989; Boesch and Boesch-Acherman 2000) underscore the behavioral flexibility of chimpanzees and suggest context-specific hunting adaptations. At Gombe, chimpanzees hunt a variety of prey, including a small percent of infant colobus monkeys; at Taï, they specialize in adult red colobus. At Gombe, the majority of hunts are by individuals; at Taï, the majority of hunts are cooperative. At Gombe, the colobus fight chimpanzees loudly and aggressively; at Taï, they flee silently. Some of these differences are apparently due to habitat structure differences between Gombe and Taï (Boesch 1994).

As several investigators have noted, the robust and successful hunting strategies of wild chimpanzees suggest the presence of this adaptation in the last common ancestor of chimpanzees and humans, and hence hunting in early hominins (Wrangham 1987; Boesch and Boesch 1989; Stanford 1996; Wrangham and Peterson 1996; Stanford 1999). Boesch's (1994) study further suggests the flexible, strategic, context-specific nature of that hunting. In addition, the omnivorous diet of chimpanzees, and especially their reliance on embedded and encased foods such as termites, ants, nuts, and honey, made accessible by tool use (i.e., extractive foraging), suggests a similar dietary breadth in the last common ancestor (Parker and Gibson 1979).

Reconstructing Subsistence Patterns in Basal Hominins (Seven to Four Million Years Ago)

Given the subsequent increasing divergence of hominins from African apes, we should expect a significant shift away from the common ancestral pattern. Some have suggested that this shift involved exploitation of new resources in the emerging grasslands as well as increased dietary breadth (Sponheimer and Lee-Thorp 2003). The diet of basal hominins likely involved intensification of some elements of that broad omnivorous pattern that were compatible with their small body and brain size. It would also be consistent with the lack of worked stone tools and prey remains associated with these fossils. Use of perishable tools and hammer stones exceeding that of chimpanzees is almost certain (Panger, Brooks, et al. 2002). Very little is known about these recently discovered creatures. All of the subsistence scenarios for early hominins including those described as "protohominids" (known as protohominins in modern taxonomy) focus on the australopithecines rather than the more recently discovered basal hominins.

It seems likely that basal hominins expanded the scope and frequency of tool use and food transport devices as part of their ecological shift into woodland habitats. Lack of worked stone tools associated with these species suggests that their hunting and/or scavenging was limited to small prey, which may have been dispatched with wooden points and clubs and unmodified stone tools. In sum, we envision basal hominins as having shifted in emphasis to greater reliance on extractive foraging on a wide variety of embedded foods in woodland habitats, but otherwise little changed from the last common ancestor (LCA) (Parker and Gibson 1979).

Subsistence Patterns of Early Human Ancestors (3.5 to 2.4 Million Years Ago)

Until the discovery of *Ardipithecus* and other basal hominins, australopithecines and related species were considered protohominins. Like their predecessors, these hominins displayed small brains and body sizes and significant sexual dimorphism in body size. They had thicker dental enamel, larger cheek teeth, and thicker lower jaws. They continued to live in woodland habitats. They probably continued to climb trees to avoid predators, though their developing upright posture and locomotion probably made them more accurate missile throwers (Dart 1925; Johanson, White, et al. 1978).

There is ample evidence they used simple worked stone tools as early as 2.5 million years ago (Semaw, Renne, et al. 1997; M. Leakey, Feibel, et al. 1995; M. Leakey, Feibel, et al. 1998; Asfaw, White, et al. 1999; M. Leakey, Spoor, et al. 2001). Based on hand morphology and behavioral continuity, it is virtually certain that they had used unmodified stone tools and other tools long before this (Panger, Brooks, et al. 2002).

Worked stone tools in association with cracked marrow bones of antelope reveal that some of these creatures had expanded their subsistence strategies to include consumption of marrow bones and brains of larger prey animals, captured through scavenging (Asfaw, White, et al. 1999). These high in protein and fat food sources are unavailable to other scavengers and hunters, except for hyenas.

Recent research on strontium/calcium ratios in the teeth of South African mammals reveals that the diets of both *A. africanus* and *Paranthropus* differed from that of chimpanzees in having a greater component of foods that use the C4 pathway of photosynthesis. These foods could have included grasses, sedges, and/or grass-eating termites, all of which are abundant in the emerging grasslands. *A. africanus* also has a larger C4 component than *Paranthropus*, which resembles baboons in its profile (Sponheimer and Lee-Thorp 2003; Sponheimer, Lee-Thorp, et al. 2005; Sponheimer, de Ruiter, et al. 2005).

Research on use wear on a large sample of animal bones associated with these early hominins in South Africa reveals wear indicative of termite foraging rather than digging for tubers (Blackwell and d'Errico 2001; Shipman 2001).

Dart's Killer Ape Scenario

In his article "The Predatory Transition from Ape to Man," Raymond Dart (1953) argues that "ape-men," represented by *Australopithecus africanus*, evolved upright posture in the service of carnivorous predation. Noting his debt to Darwin, he says erect posture

> emerged through and was consolidated by the defensive and offensive stone-throwing and club-swinging technique necessitated by attacking and killing prey from a standing position. This purposive specialization of the hands in accurate hitting & throwing . . . was the only persistent stimulus capable of transferring the body weight from the clambering knuckles and the sitting ischial tuberosities of the apes to the only suitable base for the torsional work as man's body performs. (Dart 1953, 209)

He asserts that men universally possess the inherited instinct for accuracy in hitting and hurling.

Comparing modern humans to great apes, Dart (1953, 209) notes that "living anthropoids, lacking bipedal fixity, have developed so little accuracy in wielding or hurling objects to strike other creatures . . . that they all rely on their teeth or nails . . . when struggling in close quarters." He also says "the divergence in skill between apes and men in hitting and hurling depends upon the acquisition by human beings of a brain capable of co-ordinating with hand and eye movements a series of postural body reflexes" (110). However, in contrast to Elliot-Smith, who argued that the brain was the fundamental factor behind human evolution, Dart argues that "it is absurd to imagine now that a large brain is essential to perform such berserk deeds" (205).

Middle-Period Human Ancestors (1.9 Million Years Ago to Three Hundred Thousand Years Ago)

Most anthropologists agree that *H. erectus/ergaster* marks a major transition in hominin evolution. These hominins were larger in both body and brain size and were generally believed to be less sexually dimorphic than early human ancestors. A recent find of a small female skull suggests that they may have been highly dimorphic (Spoor, Leakey, et al. 2007). They had modern trunk and limb proportions associated with more efficient bipedal locomotion. Their anterior teeth were enlarged and molar teeth were reduced in size compared to their ancestors. Unlike their ancestors, they produced biface worked stone tools (Acheulean tools), ate larger prey, probably controlled fire, and moved out of Africa for the first time.

Stable isotope studies reveal that the appearance of *H. erectus/ergaster* corresponds to a major environmental change to more open, grassy ecosystems

Based on finds at Makapansgat associated with australopithecines, Dart argues that these ape-men hunted large extinct bears, horses, elephants, and giraffes using the long bones of antelope as clubs, making it clear that the animals slain by the australopithecines were "huge and active." Moreover, he says that the large bones of animals left by man-apes at Makapansgat have been split and crushed to extract the marrow. He concludes that "the carnivorous dietary habits of the South African man-apes . . . have demonstrated the direct relationship which exists between the carnivorous habit and acquiring the upright posture" (1953, 205).

Dart enumerates the widespread occurrence among living men of flesh eating, including cannibalism: "The loathsome cruelty of mankind to man forms one of his inescapable characteristic and differentiative features; and it is explicable only in terms of his carnivorous and cannibalistic origin" (1953, 207). Although Dart describes the precursor to australopithecines as "ancestral bipedal brachiatiors," he supports his view with an ecological argument about the implications for human evolution of the carnivorous habits of terrestrial baboons in South Africa.

Dart elaborates on these themes in his later book *Adventures with the Missing Link*, in which he also details his ideas about ape-man's use of animal bones, horns, and teeth as tools for hunting and butchering, that is, his "osteodontokeratic culture."

at about 1.8 million years ago. This change entailed increases in grasses, underground storage organs (USO), and tuber- and grass-feeding animals (Lee-Thorp and Sponheimer 2007).

Subsistence of Recent Human Ancestors: Archaic and Modern *Homo sapiens*

About three hundred thousand years ago, there was a major transition from Late (Acheulean) to Middle Stone Age technology as flake tools and hafted implements replaced hand axes and cleavers (McBrearty and Brooks 2000). The following transition from the Middle to the Late Stone Age occurred sometime between two hundred thousand and thirty thousand years ago. The traditional view is of the sudden revolutionary emergence of modern behavior about thirty thousand years ago in the Upper Paleolithic in Europe. The alternative view is of a gradual transition into modern behavior during the Middle Stone Age (MSA) in Africa (McBrearty and Brooks 2000).

In Europe, the subsequent appearance of anatomically modern *H. sapiens* about forty thousand years ago coincided with revolutionary changes in technology. The so-called Upper Paleolithic technology of the immigrants replaced the Middle Paleolithic (Mousterian) technology of Neanderthals. It included production of geometric microliths for composite (hafted) tools, prismatic blade cores, pressure-flaked projectile points, heat treatment of raw materials to improve their tool making, spear throwers, bow and arrows and other mechanical devices, bone and antler ivory technology, bone needles, and both ground stone artifacts and ceramic vessels. Finally, and most famously, they made representational art, both mobile and fixed on rocks and caves (Toth and Schick 1993).

They developed new technologies for food harvesting, transport, preparation (e.g., leaching poisons, drying meat and vegetables, grinding seeds and nuts), and storage. Certainly seeds had become important foods long before this (C. Jolly 1972). In addition to these technological advances, they expanded their range and diversity of foods (Flannery 1969; Stiner 2001). They also began moving into new habitats, specializing in exploitation of different ecosystems, including arctic and marine habitats. Hunters began to specialize in prey. They began to anticipate and harvest annual migrations of game, forming large seasonal aggregates to do so (Conkey 1980). Alliances between groups facilitated exchanges, which became an important means for evening out seasonal and annual variations in resources (Gamble 1976).

In contrast, in Africa the transition between the Middle Stone Age and the Late Stone Age was much more gradual. According to McBrearty and Brooks, the MSA archaeological sites of archaic *H. sapiens* in Africa suggest exploitation of desert and forest as well as other challenging habitats. They also suggest that they were competent hunters and fishers who planned their settlements to harvest seasonal aggregations of game and fish. They used arrows in hunting large game and grindstones for preparing plant foods. Well-differentiated regional styles in stone points in Africa predate those in Europe by tens of thousands of years. Likewise, composite projectiles were present in Africa sixty-five thousand years ago, approximately thirty thousand years earlier than in Europe. Worked bone artifacts are associated with other MSA artifacts, as well as ornaments including perforated shell beads and grindstones with traces of pigment. Overall, McBrearty and Brooks argue that "distinct el-

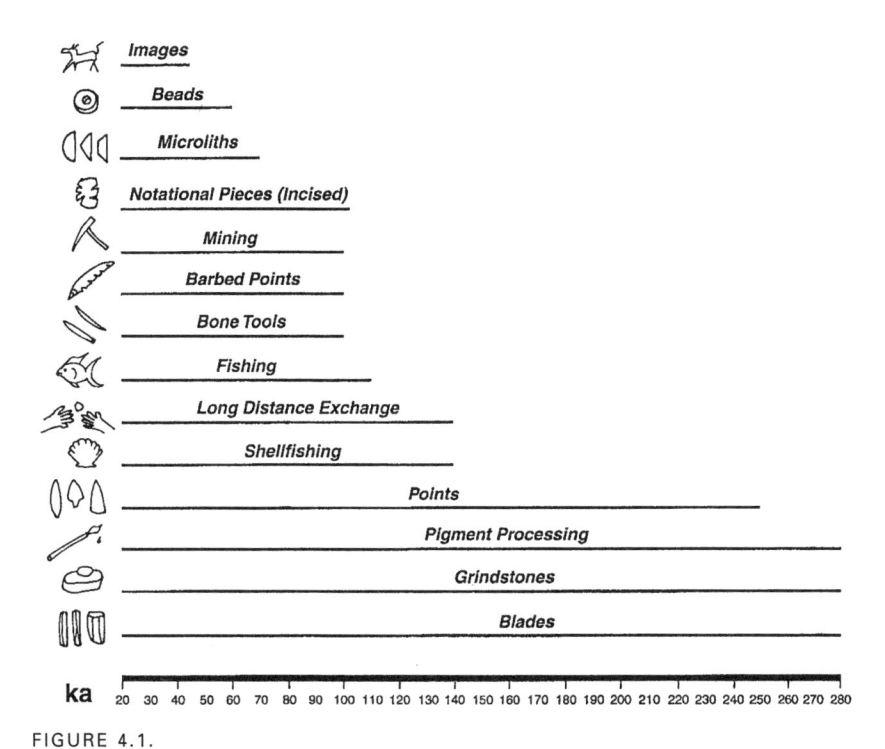

FIGURE 4.1.
Behavioral innovations of the Middle Stone Age in Africa.
Source: McBrearty, Sally, and Alison S. Brooks. 2000. "The revolution that wasn't: A new interpretation of the origin of modern human behavior." Journal of Human Evolution 39 (5): 111. Copyright © 2000 Elsevier Limited. Reprinted with permission from Elsevier.

ements of the social, economic, and subsistence bases changed at different rates and appeared at different times and places" (McBrearty and Brooks 2000, 458). See figure 4.1 for a graphic depiction.

By about ten thousand years ago in the Middle East, probably stimulated by climate change favoring grasses, modern humans had begun to domesticate seeds and animals (Richerson and Boyd 2000; Richerson and Boyd 2003). Population densities increased dramatically. Agriculture brought innovations in architecture and social organization, including monopolization of food and other resources by powerful families leading to increased social stratification and increased warfare among groups. Food became a means for status display and intimidation (Hayden 1994).

Washburn and Lancaster's Hunting Scenario

In their article "The Evolution of Hunting," Washburn and Lancaster (1968, 228) say that "the biology, psychology, and customs that separate us from the apes—all these we owe to the hunters of past time." They stress that human hunting is more than a mode of subsistence. It is "a set of ways of life. It involves divisions of labor between male and female, sharing according to custom, cooperation among males, planning, knowledge of many species, and large areas of technical skill" (217).

Specifically, they argue that the family arose from habitual sharing between male hunters and female gatherers and their offspring. The long-term cooperative associations within the family favored the development of the incest taboo as a means for avoiding conflict between sexual, economic, and authority roles, and reducing the number of mouths to feed.

On the basis of comparative data on living hunter-gatherers, they argue that early hunter-gatherers would have had large territories and small populations with low growth rates. Consequently, they would have had unbalanced sex ratios, which would, in turn, favor exogamy and exchange of mates and the beginning of language for planning these exchanges.

Washburn and Lancaster argue that hunting has dominated human history since the time of *Homo erectus* at about six hundred thousand years ago, and perhaps *Australopithecus* may have hunted, albeit less efficiently, as long as three million years ago. Noting the technical skill involved in producing symmetrical Acheulean biface tools, and the (three times) larger cortex of humans as compared to australopithecines, they emphasize that "success of tools has exerted a great influence on the evolution of the brain" (1968, 221).

They also emphasize that the hunter's worldview, which entails a lack of control over killing conspecifics as well as game, is based on pleasure in killing. Citing various ethnographic studies, Washburn and Lancaster also argue that "the skills for killing and the pleasure of killing are normally developed in play" (1968, 223). Moreover, they note that "men enjoy killing, and these activities are continued as sport, even when they are no longer economically necessary" (222).

Finally, they emphasize the adaptability of the hunting and gathering way of life, which allowed humans to occupy most of the earth. They note that this adaptability was already apparent half a million years ago in the geographic range of *H. erectus*.

Before leaving our survey of ancestral diets, it is important to note that all the stages or periods of human evolution were preceded by and occurred in the context of major climate changes. Various new technologies (for example, cores from deep-sea sediments and ice cores) have revealed tumultuous and deteriorating climates beginning fourteen million years ago in the Mesozoic era, and continuing and intensifying conditions of rapid and chaotic changes in ecosystems during the past 2.5 million years in the Pleistocene (Richerson and Boyd 2000). As noted above, paleoecological constructions in sub-Saharan Africa reveal that grasslands first emerged six to eight million years ago, and became a major component of eastern and southern African vegetation 1.8 million years ago (Lee-Thorp and Sponheimer 2007).

SCENARIOS FOR SUBSISTENCE PATTERNS OF HUMAN ANCESTORS
Subsistence Scenarios for Early Human Ancestors

Scenario builders differ widely in the subsistence patterns they attribute to early hominins (usually australopithecines). Hunting (Darwin 1859; Dart 1953; Washburn and Lancaster 1968; Wrangham and Peterson 1996; Stanford 1999), scavenging (Bartholomew and Birdsell 1953; Szalay 1976), gathering vegetable foods (Bartholomew and Birdsell 1953; Tanner and Zihlman 1976), seed eating (C. Jolly 1972), excavating roots and tubers (O'Connell, Hawkes, et al. 1999; Wrangham, Jones, et al. 1999), and extractive foraging on animal and vegetable foods (Parker and Gibson 1979) have all been suggested as prime movers in the evolution of hominins.

The best direct archeological evidence is for scavenging, but there is good reason to believe that these hominins engaged in all of the suggested subsistence strategies. First, given that all other mammalian scavengers also engage in hunting, it is likely that australopithecines did, too (Schaller and Lowther 1969). Continuities with the hunting behavior of both chimpanzees and human foragers also argue for this (Stanford 1999).

Given the uncertainty of hunting, however, neither human nor chimpanzee hunters can survive on meat alone. With the exception of Arctic peoples, human foragers obtain the bulk of their subsistence from gathered foods (Lee 1979). Moreover, the teeth of australopithecines reveal a broad-spectrum vegetarian diet (Sponheimer and Lee-Thorp 1999; Van der Merwe, Thackeray, et al. 2003). Therefore, consumption of vegetable foods including underground storage organs (USOs)—tubers—is virtually certain. USO eating seems likely given the abundance of this resource in woodland habitats and lack of competitors, given that USOs are inaccessible to non–tool users. Taphonomic evidence reveals tool use in termite foraging (Shipman 2001).

Each dietary category implies different behaviors. First of all, these scenarios imply different kinds of tool use. Eating USOs implies use of large digging sticks. Extracting such other embedded foods as nuts and hard-shelled fruits implies hammer stones, excavating ants and termites implies termite-fishing wands, honey harvesting implies use of small sticks. Eating these and other gathered foods such as seeds and eggs suggests use of containers for collecting and transporting them. Scavenging implies throwing aimed missiles at competing scavengers. Consumption of small nesting and fossorial animals suggests throwing and clubbing. Both of these activities would entail bipedal stance and locomotion. Second, the USO scenarios imply larger molar teeth with thicker enamel for coping with fibrous foods and grit. Gnawing on bones may have favored reduced canine teeth (Szalay 1976), which are also characteristic of these creatures. Chewing grains and seeds may have favored canine tooth reduction, larger molars, and rotary jaw motions (C. Jolly 1972).

Tanner and Zihlman's Scenario for Hominid* Subsistence

Tanner and Zihlman (1976) contrast their scenario with earlier hunting scenarios focusing on the role of males, which are "incomplete and misleading" insofar as they neglect the role of women. Their first article focuses on the transition from apes to humans occurring about four to six million years ago: "Specifically, we hypothesize the development of gathering as a dietary specialization of savanna living, promoted by natural selection of appropriate tool using and bipedal behavior. We suggest how this inter-relates with the roles of maternal socialization in kin selection and of female choice in sexual selection. We emphasize the connections among savanna living, technology, diet, social organization, and selective processes to account for the transition from a primate ancestor to the emergent human species" (586).

They suggest that the common ancestor of apes and humans may have been similar to chimpanzees. They envision a shift away from fruit eating in this ancestor to an omnivorous diet in the transitional form coincident

---

*Hominids are now known as hominins.

with a shift from the forest to the forest fringes. They argue that "bipedalism freed the hands for tool use in obtaining and carrying food, enhanced visibility, and increased the effectiveness of displays against potential predators" (Tanner and Zihlman 1976, 587). Also, canine teeth were reduced in size, which increased chewing efficiency and increased sociability.

After reviewing concepts of natural and sexual selection, Tanner and Zihlman discuss new evidence for the close genetic relationship between chimpanzees and humans. They argue that this confers new relevance on data on contexts of chimpanzee bipedalism: carrying food, supporting infants, displaying with objects, and looking over grass. They also note that chimpanzees do the following things previously thought to be unique to humans: "They make and use tools for feeding, grooming, and investigation, and nests for sleeping. They catch, kill, and eat small animals. They share food" (1976, 591). Nevertheless, they emphasize the predominantly plant base of the diet, and the greater incidence of tool use among female chimpanzees. They also focus on the heavy maternal investment in offspring and lifelong ties between mothers and offspring.

Their model of the transitional population moving into a new savanna habitat focuses on increasing dietary emphasis on seeds, nuts, insects, roots, eggs, and meat. Tanner and Zihlman propose that females innovated by gathering these foods with digging sticks and transporting them in containers to safe locations. They also suggest that mothers carried nonclinging offspring in slings and scared predators off with thrown objects and noisy displays rather than climbing trees. They envision a flexible social structure based on stable mother-offspring subunits as part of small groups separating and congregating: "When we speak of 'kin' for the transitional hominids, we refer to this loose genealogical set of mother, offspring, siblings, mother's siblings and possibly mother sister's offspring" (1976, 605). Finally, they argue that females were sexually assertive creatures engaging in female choice of mates, preferring males who were friendly and nonthreatening: "The picture then is of bipedal, tool-using, food-sharing and sociable mothers choosing to copulate with males also possessing these traits" (606).

In part two of this article, Zihlman (1978) continues the feminist theme. She begins with a characterization of the last common ancestor of African apes and humans: "About 5 million years ago forest-ranging, knuckle-walking apes—very much like chimpanzees—evolved through the process of natural selection into the earliest humans, the hominids, who walked upright on two legs, used tools, and lived and gathered food on the African savanna" (4).

She follows with a description of "our early ancestor, Australopithecus" in their mosaic habitat of grasslands, low bush, and riverine forests, characterizing them as omnivorous gatherers who continued the chimpanzee pattern of preying on small animals and using and making perishable tools. Noting the small canine teeth and body size of these creatures, she suggests that they relied on safety in numbers and aggressive throwing displays to ward off predators. She suggests that the hominid way of life required a longer period of socialization and hence longer dependency of offspring.

Zihlman suggests that mate choice by females lacking specific estrus periods favored less aggressive males. She proposes that "among the australopithecines the economic units were primarily the smaller kin groups that shared plant and animals foods and cared for the young. Sexual behavior and the 'reproductive units' occurred within the larger associations of unrelated individuals who came together in their kin groups and food and water sources and sleeping places. These two units become linked much later in time" (1978, 11).

In her discussion of the evolution of *Homo erectus*, Zihlman argues that these creatures continued to rely on flexible and opportunistic methods of subsistence that are obscured by the label "big game hunters." She traces the large home ranges and home bases of *H. erectus* to gathering. She also argues that early hunting and gathering technologies may have been little different and that tools for hunting may have developed from gathering tools: for example, she says that later use of spears may have arisen through generalization of the digging stick.

Subsistence Scenarios for Middle-Period Human Ancestors

Three major subsistence adaptations have been proposed for *H. ergaster/erectus*: hunting large game (Washburn and Lancaster 1968), scavenging (Shipman 1986; Blumenschine 1987), and harvesting and cooking USOs (O'Connell, Hawkes, et al. 1999; Wrangham, Jones, et al. 1999). The hunting scenario goes back to Darwin (1868) and to Dart (1953) who placed it among australopithecines. Likewise, the scavenging hypothesis was originally proposed for *Australopithecus*. More recent investigators have applied all three hypotheses to *Homo erectus*.

The hunting model has been the most persistent and popular model. It fell out of favor in the seventies and eighties, first with the seed-eating hypothesis, and then with the rise of feminism and the emphasis on gathering that came out of studies of living hunter-gatherers. These studies indicated that some foragers took most of their calories from foods gathered by females (Bicchieri 1972; Teleki 1975). Subsequent studies raised questions about the model of bonding through food sharing, suggesting instead that the best hunters used their prowess to gain copulations with many females. Conversely, more recent studies of chimpanzee hunting have stimulated reexamination of this scenario (Boesch 1994; Stanford 1998; Stanford 1999; Wrangham 1999).

Washburn and Lancaster (1968) argue that beginning around six hundred thousand years ago *H. erectus* depended upon hunting and gathering and that their social life was shaped by this subsistence mode. Specifically, they had large territories, small populations, and a division of labor with sharing between males and females.

Subsequent investigators have elaborated these patterns in light of fossil and archaeological data. Walker and Shipman (1996), for example, note that *H. erectus* dentition indicates a meat diet, as does its smaller but longer digestive tract. Walker argues that the movement of this species out of Africa is consistent with his observation that herbivores-turned-carnivore have larger home ranges as well as smaller population densities. He also notes that they have new means for removing meat from carcasses (tools) and are more social than their herbivorous ancestors, as expected. Finally, he discusses the case of hypervitaminosis that killed the fossil *H. erectus* known as 1808, probably as a result of eating the liver of carnivores.

The occurrence of long bones cracked open for marrow and cut marks on bones of herbivorous prey clearly indicates butchery and meat consumption, but does not clearly distinguish scavenging from hunting. Hunting is indicated by disarticulation of bones seen in cut marks near joints for transport, skinning marks that are more frequent than meat removal marks, and cut marks overlain by carnivore tooth marks. Scavenging is indicated by the converse: lack of disarticulation marks, processing marks on non-meat-bearing areas, and cut marks over carnivores' teeth marks. Based on these predictions, Shipman (1986) concludes that cut marks on herbivore bones from Bed I Olduvai Gorge support the scavenging hypothesis.

Proponents of scavenging have done two other kinds of supporting studies. First, Blumenschine (1987) has done field studies of the availability of abandoned carcasses (completeness and persistence by prey size) in East African game parks according to habitat and season. He discovered that lion kills of medium-sized herbivores were most abundant, complete, and persistent. Competing scavengers always leave bone marrow and brain. Hyenas eat both these organs in open environments, but rarely enter riparian woodland environments. Following this, he compared these data with data on Plio-Pleistocene herbivores, carnivores, and scavengers at Olduvai Gorge and East Lake Turkana, two hominin fossil sites. He concluded that marrow and head contents from medium-size and large carcasses may have been available more frequently in the Plio-Pleistocene than they are today and that flesh from these creatures was also more available then.

Second, Shipman (1986), based on comparative data on birds and mammals, has proposed several signs for recognizing scavenging in the fossil record: (1) locomotor endurance, (2) adaptation for locating carcasses, (3) means for dealing with competitors either through bullying or sneaking, and (4) using a reliable alternative food source. Based on her study of the fossil record of Bed I Olduvai Gorge in Tanzania, she concludes that all of these adaptations were present in the hominins of that location. Specifically, bipedal locomotion served the need for locomotor persistence and for carcass location. Tree climbing and tool use for speedy removal of tissue served as competitive measures, and fruits and vegetable foods served as a reliable alternative food.

Nevertheless, like Schaller and Lowther (1969), Shipman (1986) notes that no mammalian scavenger relies exclusively on this mode of subsistence. She proposes a *hunting scavenging continuum*. Likewise, proponents of the USO scenario suggest that *H. erectus* hunted as well as gathered. According to Wrangham and colleagues (1999), *H. erectus* depended heavily on cooked USOs because they provided an abundant, reliable supply of energy. This energy was rendered more digestible and palatable by cooking, which also made it more concentrated and valuable and hence subject to theft. The tuber-cooking scenario relies on the availability of fire. Although archaeological remains of fire are disputed, Rowlett (1999) argues that good evidence is available from several sites in Africa.

Szalay's Scavenging Hypothesis

Frederick Szalay (1976) proposes a model of early hominid (older term for hominin) scavenging based on an analysis of australopithecine dentition. He opposes his model to the gelada baboon seed-eating hypothesis Clifford Jolly (1972) proposed to explain canine reduction and bunodont molars and manual dexterity in hominids. Szalay argues that the "unique occurrence of no sexual dimorphism in canines but its retention in cranial and somatic characters and in general size requires some special explanation" (420). The special explanation he proposes is positive selection for use of the anterior dentition resulting in *incisivation* of the canines with "interdependence of the canine-incisor row" (420). He emphasizes transformation rather than size reduction, noting that "the hominid canine is not vestigial but has become transformed, with accentuated mesiodistal cutting edges, into an incisiform tooth, and has incorporated, in an apparently highly exacting manner, into the continuous apical edges of the canine incisor row" (425).

In Szalay's (1976, 428) view, hominid incisors are adapted for grasping and tearing meat from bones: "Strong, vertically implanted incisors plus incisiform canines become the tools which grasp and tear meat, tendons, and fascia. Selection for this role would have been intense, as without good tools, much meat could not be ripped off carcasses left over by larger predators." Likewise, he argues that strong mandibles and thick molar enamel would be advantageous, "either for withstanding the abrasive effects of cracking ribs, metapodials, joints or increased longevity or both" (428).

Turning to the habitats of early hominids, Szalay suggests that during the African Miocene transition from forests to open woodland and savanna, these creatures may have stayed closer to the woodland and forest.

Szalay's proposal is accompanied by a detailed methodological and anatomical critique of Jolly's seed-eating hypothesis. He criticizes Jolly for failing to do a functional analysis of dentition, and for ignoring or glossing over significant differences in the dental adaptations of gelada baboons and hominids.

Wrangham, Jones, and colleagues (1999) suggest that increased reliance on hunting was made energetically feasible by the consistent availability of this foodstuff. O'Connell, Hawkes, and colleagues (1999) also argue that USOs were a staple food for *H. erectus*. Based on ethnographic data on hunter-gatherers,

Wrangham, Jones, et al.'s Subsistence Scenario for Origins of *Australopithecus* and *Homo erectus*

Wrangham and colleagues (Wrangham, Jones, et al. 1999) propose that cooking of underground storage organs (USOs—roots and tubers) by *H. erectus* was the crucial new subsistence adaptation leading to several derived characters including reduced tooth and gut size, reduced sexual dimorphism, increased absolute body size, and increased female sexual attractiveness: "We propose that the new relationship between foraging and social competition caused by cooking [a producer-scrounger system] led to pressure on females to form protective bonds with males" (568).

The authors note that the increased energy expenditure of *H. erectus* relative to australopithecines is usually attributed to increased meat consumption. In contrast they argue that these hominids (now called hominins) continued to depend on plant foods during times of food shortage, despite meat consumption. Wrangham, Jones, et al. (1999) say this is the case because whereas hunting is always chancy, cooking consistently renders foods more digestible, thereby enlarging and improving the diet. It also concentrates food into high value packets subject to theft. Moreover, they suggest that increased meat consumption in later hominids was secondary to the increased energy intake made possible by cooking. They acknowledge uncertainties about archaeological evidence for use of fire in cooking. They note that control of fire would

they argue that hunting is unreliable, and that males hunt for prestige and for greater access to females. Their scenarios for the life history and mating consequences of this subsistence strategy also differ from those proposed by the hunting scenario (see chapter 6 on life history).

Given the apparent history of omnivorous diets in fossil and living hominins, we suspect that *H. erectus* engaged in all of the proposed subsistence activities. Based on current evidence, the two most significant changes associated with the diet of these creatures seem to be the increased size of their prey (both hunted and/or scavenged), and the increased proportion of meat in their diet. This shift was accompanied by and depended on increased sophistication of tool production and use, primarily for bone cracking and butchery, both extensions of extractive foraging. Harvesting and preparing animal

have brought many benefits including protection from predators, freeing hominids from protection of trees, and increased scope of reciprocity.

Rather than attributing male-female bonding to provisioning of females by male hunters, Wrangham, Jones, et al. (1999) suggest that bonding may have arisen as females formed alliances with dominant males to protect their concentrated, high-value foods from theft by other males. They attribute reduced sexual dimorphism to increased number of mating days between births arising from increased female attractiveness.

They also propose that australopithecines used USOs as fallback foods during periods of food scarcity. They argue that this was responsible for the derived features of thickened molar enamel and large molar surfaces in these hominids. They argue that use of these foods allowed australopithecines to colonize the savanna with the aid of digging sticks.

Wrangham, Jones, et al. (1999) present several kinds of data to support their proposals. First, food- and water-storing USOs occur in high densities and high species diversity in dry woodland savannas and drier habitats from Ethiopia and Chad to South Africa. Second, fossil mole rats, drier habitat species that depend on USOs, are found in the same habitats as early fossil hominids. Third, these USOs generally occur below 10 cm, which makes them inaccessible to non–tool-using mammalian taxa, including baboons, and even to tool-using chimpanzees. Fourth, living African and other tropical hunter-gatherers rely heavily on USOs during periods of food scarcity.

---

and vegetable foods was only one arena for technology; construction of shelters and clothing was another, which must have made expansion into Asia and Eurasia possible for the first time.

Both archaeological and circumstantial evidence suggest that production and use of fire arose at this time. Fire must have been crucial for maintaining adequate warmth as well as for scaring away predators and competitors. Use of fire for cooking food would have enlarged the scope of edible foods, enhanced their value and palatability. Cooking food must have been a major change in food preparation, which ultimately brought reduction of molar teeth (Brace, cited in Wrangham, Jones, et al. 1999). Although it is unclear when this process began (Wrangham, Jones, et al. 1999), certainly it had occurred by two hundred thousand years ago, when archaic *H. sapiens* had spread throughout the Old World. These creatures lived in harsh winter conditions and hunted big game. Their Mousterian and related Middle Paleolithic tool kits were more complex and diversified than the Acheulean tools of *H. erectus*.

Subsistence Scenarios for Recent Human Ancestors

Perhaps because the archaeological data for this period are relatively abundant, scenario builders have not focused on it. Archaeological and ethnographic evidence suggests increasing specialization of hunting, foraging, and food storage and preparation. These include planting, harvesting, winnowing, grinding, leaching, mixing, and underground storage, as well as cooking. Use of water to leach poisonous chemicals out of such foods as cassava and acorns is one obvious example. Poisons so extracted are used for hunting and other purposes (Bicchieri 1972). Similar processes can be used to extract medicinal, cosmetic, and hallucinogenic substances from plants and animals.

Extracting, processing, and planting seeds from grain and other vegetable crops was crucial in the emergence of agriculture. Use of simple mechanical devices such as wheels to extract mechanical energy from flowing water or wind is another vital application of extractive technology. Melting mineral ores over fire to extract valuable metals was crucial in the emergence of metallurgy (Mason 1966 [1895]; Oswalt 1973; Vogel 1997). Later, extraction of petroleum-based chemical energy from ground reserves became crucial to the industrial revolution. A good case can be made that extractive technologies have played a crucial role at every stage of hominin evolution and continue to play a vital role in modern life.

CONCLUSIONS ON HOMININ SUBSISTENCE

Focusing on continuities as a bridge to new adaptations seems to us to be the best strategy for evolutionary reconstruction. Therefore, we agree with Bartholomew and Birdsell (1953) that early human ancestors, like the LCA and basal hominins, were omnivores that employed all of the strategies—small game hunting, scavenging, and gathering—especially during seasons of food scarcity in woodland habitats. USOs apparently offered an abundant new source of food to those capable of using large digging sticks (Wrangham, Jones, et al. 1999), and of mechanically processing these foods (McGrew 1999). Likewise, new grasslands provided abundant invertebrate food—termites—for tool users. Also, harvesting of nuts, seeds, honey, eggs, small fossorial animals, and ants must have continued (they are commonly used by chimpanzees and living foragers). All of these foraging strategies, including breaking open bones for marrow, are forms of *extractive foraging with tools*. Hominins are specialists in this generalist strategy.

It seems to us that all of these subsistence models describe important elements of the ongoing elaboration and extension of hominin diets. In many cases, however, the timing of these elaborations is difficult to determine. Modern paleoanthropologists, for example, have displaced the hunting hypothesis from australopithecines to *Homo sapiens*. Archaic *H. sapiens* with Middle Stone Age and Mousterian tool cultures, including a variety of end and side scrapers and points generated by the Levallois prepared-core technique, may have depended on hunting as well as scavenging of large game. Only with the emergence of more complex elaborated Late Stone Age tool cultures with blades and hafted tools, however, does there seem to be unambiguous evidence for large-scale dependence on hunting of large game animals and controlled use of fire (Klein 1999; McBrearty and Brooks 2000).

We see both continuity and transformation in subsistence patterns from stage to stage of hominin evolution. First, the eclectic omnivorous diets of hominins (based on a variety of seasonal and embedded vegetable and animal foods supplemented by small prey) increased as new grassland habitats expanded. Second, extensions and elaborations of extractive foraging technology during sequential stages of hominin evolution expanded the quantity and availability of these foods:

1. As grasslands first appeared, basal hominins apparently increased their dependence on various embedded foods accessible only by tool use;
2. Extensions to worked stone tools as hammer stones by early hominins opened up opportunities for extracting bone marrow and head contents of larger prey they scavenged to supplement gathered foods;
3. As grasslands expanded, elaborations of stone tools by middle-period hominins opened up opportunities for hunting and butchering large- and medium-size game, again to supplement gathered foods;
4. In *H. sapiens*, increasing elaboration of projectile tools, supplemented by knowledge of animal behavior, led to successful large-scale hunting and exploitation of new habitats and food sources.

At some point, cooking and other food processing techniques led to further expansion of diets to include previously indigestible plants (USOs and grains) and animals. Alliances and trade relations further expanded food availability. Finally, domestication of plants and animals led to settled communities and hastened population growth.

Extractive Foraging Continuum

In contrast to other hominin subsistence hypotheses, the extractive foraging model emphasizes evolutionary continuities and new elaborations in both subsistence technology and cognition (see chapter 9). These continuities are revealed by the preceding review of extractive foraging strategies in chimpanzees and basal, early, middle, and recent hominin species. As new habitats arose, each new adaptive radiation of hominins elaborated on the extractive technologies of its predecessors.

We believe that all of the foregoing subsistence modes—foraging on termites, extracting marrow from scavenged bones, extracting meat through butchery of large game, and extracting underground storage organs through excavation—fall along an evolutionary continuum of tool-mediated extractive foraging strategies. In addition, we believe that tool-mediated extractive foraging opened the potential for entering a series of new adaptive zones in hominins under succeeding ecological challenges.

Moreover, because extractive foraging with tools is a shared derived character in the hominoid clade, it meets the *phyletic criterion* for an adaptation. Because it also evolved independently in cebus monkeys (Parker and Gibson 1977), it meets the *correlation criterion* for adaptation (see chapter 2).

5

Origins of Bipedalism

Man alone has become a biped; and we can, I think, partly see how he has come to assume his erect attitude, which is one of his most conspicuous characters. Man could not have attained his present dominant position in the world without the use of his hands. . . . The hands and arms could hardly have become perfect enough to have manufactured weapons, or have hurled stones and spears with a true aim, as long as they were habitually used for locomotion and for supporting the whole weight of the body.

—*Darwin 1871, 434*

Bipedal standing and walking is the defining characteristic of our human lineage, and one of the earliest derived characteristics to evolve. Following Darwin, the origins and adaptive significance of bipedalism have been the focus of continuing speculation. Many scenarios have been proposed, including tool use, carrying, aggressive displays, and aimed throwing. So far no consensus has been reached beyond the generalization that bipedalism freed the hands for uses other than locomotion.

In Darwin (1871) and T. H. Huxley's (1959) day, few fossil apes or hominins were known. In line with their knowledge of comparative anatomy, however, both men knew that African apes were the closest relatives of humans. They also knew that these apes engaged in both arboreal and terrestrial locomotion, including knuckle walking. Some later investigators interested in the origins of bipedalism apparently lost sight of the terrestrial component of ape locomotion and focused on their arboreal brachiation in the last common ancestor.

Sir Arthur Keith was reportedly the first anthropologist to model the locomotor mode that preceded bipedalism in human evolution. Based on anatomical similarities between humans and great apes, he postulated an upright climbing mode of locomotion consistent with the belief that apes were strictly arboreal (Richmond, Begun, et al. 2001, 71). According to Landau (1991), Keith believed that human stock evolved from a giant arboreal primate before becoming plantigrade. Keith then changed to espouse the alternative brachiationist model of Gregory and other anthropologists who believed that humans branched off a gibbon-like ancestor. Other anthropologists, including Wood Jones, Le Gros Clark, and Strauss, believed in an even more ancient split between humans and apes before apes became brachiators (Richmond, Begun, et al. 2001). Reflecting their non-Darwinian biases, none of these early investigators focused on the adaptive significance of locomotion (Bowler 1986; Landau 1991).

In the 1960s, when molecular data demonstrated that humans shared a recent common ancestor with African apes (Goodman 1963a, 1963b; Sarich 1968), Sherwood Washburn proposed that the last common ancestor of African apes and humans was a knuckle walker. He based this hypothesis on the occurrence of this form of terrestrial locomotion in all three living African apes, and on the hairless surface of human metacarpal knuckles (Washburn 1967). His student Russell Tuttle, an expert in primate functional anatomy, particularly knuckle walking, opposed this perspective, espousing a gibbon model (Tuttle 1969). Recently, however, Tuttle said that a knuckle-walking common ancestor is one possibility, though he awaits further experimental, genetic, and fossil data to resolve the issue (Tuttle, Hallgrimsson, et al. 1999). In recent years, several anthropologists have raised the question of knuckle walking (Corruccini 1978; Shea and Inouye 1993; Stokstad 2000; Lovejoy, Heiple, et al. 2001). The matter remains unsettled but recent fossil finds provide more evidence in favor of a knuckle-walking ancestor.

LOCOMOTOR PATTERNS OF HUMAN ANCESTORS
Locomotion of the Last Common Ancestor

Comparative anatomical evidence and the behavior of living African apes suggest that the last common ancestor (LCA) of apes and humans used a variety of locomotor patterns (Rose 1991), including climbing by grasping, under-branch suspension, and knuckle-walking quadrupedalism (Coruccini 1978; Richmond 2000; Coruccini and McHenry 2001; Richmond, Begun, et al. 2001). Underbranch suspensory locomotion (brachiation) involves hanging and swinging the body under branches using elongated arms, highly mobile shoulder, elbow, and wrist joints, and elongated hooklike fingers. The lesser apes, gibbons and siamangs, which are specialized for this mode of locomotion, also have shortened trunks, legs, and reduced tails. See figure 5.1 for a depiction of brachiation anatomy.

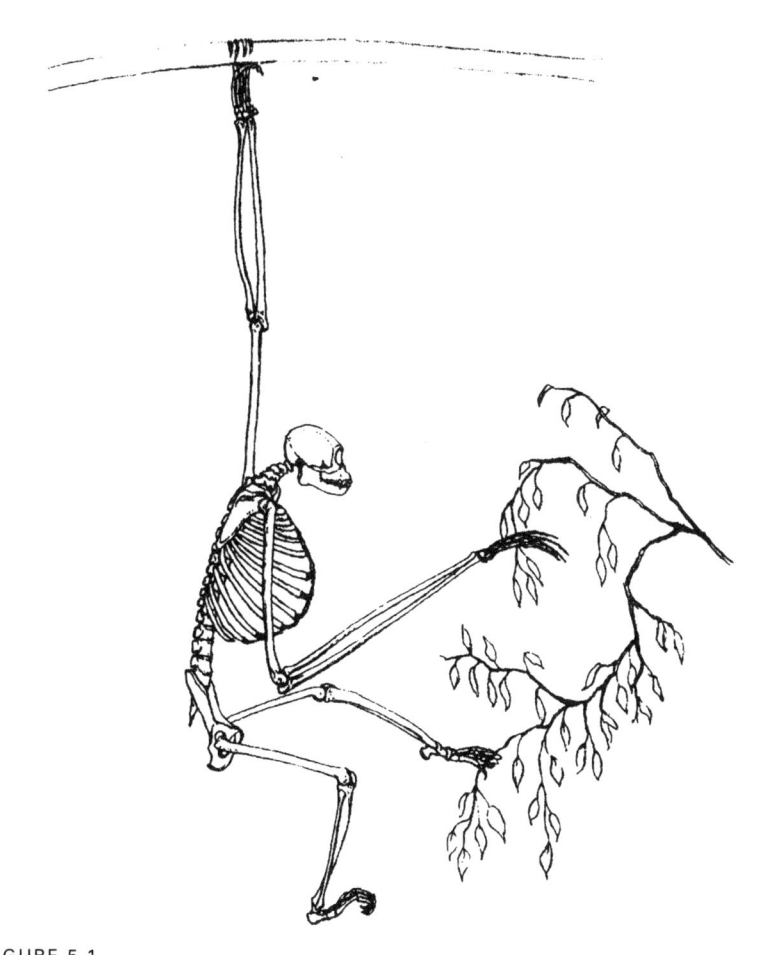

FIGURE 5.1.
Anatomy of a brachiator.
Source: Drawing originally published in Fleagle, J. G. 1999. *Primate adaptation and evolution.* New York: Academic Press, p. 240. Copyright © Academic Press 1999. Reprinted with permission.

African apes also engage in suspensory locomotion, but owing to their heavier bodies and longer legs, they spend considerable time walking on the ground bearing some of the weight of their upper bodies on the backs of their three middle fingers. Knuckle walking involves flexing the tips of the fingers and resting the forelimbs on the back surfaces of the middle fingers with either alternating arms or with both arms used as crutches. These knuckles are padded, and the wrist and hand bones and muscles keep the wrists in a rigid position during movement (Tuttle, Hallgrimsson, et al. 1999). Although knuckle walking is a form of quadrupedalism, body weight is unevenly distributed on the four limbs; the greater weight is shifted to the hind limbs during this activity. Knuckle walking comprises up to 85 percent of the locomotor repertoire of chimpanzees (Zihlman 1999). See figure 5.2 for a depiction of knuckle-walking anatomy.

Rose (1991, 39) notes that humans are unique among primates in having a locomotor repertoire dominated by a single activity. This is in contrast to other pri-

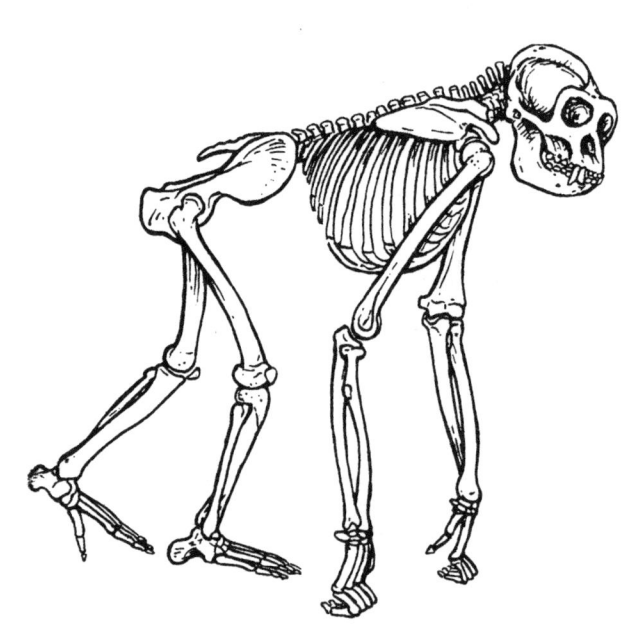

FIGURE 5.2.
Anatomy of a knuckle walker.
Source: Drawing originally published in Fleagle, J. G. 1999. *Primate adaptation and evolution.* New York: Academic Press, p. 247. Copyright © Academic Press 1999. Reprinted with permission.

mate species that have varied locomotor patterns (e.g., climbing, swinging, and quadrupedal walking), and whose bodily anatomy must "represent a biomechanical compromise biased toward the performance of the most important activities" (Rose 1991, 39). Therefore, Rose argues, we should expect that in its initial stages, bipedalism would have been added to such other locomotor activities as climbing and quadrupedalism in the ancestral human's repertoire.

After reviewing the anatomy of living and fossil great apes, Richmond, Begun, et al. (2001, 95) conclude that "the comparative anatomy of extant anthropoids and fossil Homininae suggest that humans evolved from a knuckle-walking ancestor. Upper limbs and trunk features are consistent with some form of climbing, whereas other upper limb features are consistent with terrestriality and in a variety of ways, knuckle-walking."

The most parsimonious explanation for the distribution of these features among fossil and living hominin species that display them—African apes and early human ancestors—is that they retain these traits from a knuckle-walking common ancestor (see Richmond and Strait, 2000). A knuckle-walking common ancestor of African great apes and humans is consistent with a *Pan-Homo* clade within the Homininae (otherwise, knuckle walking had to have evolved twice, once in gorillas and once in chimpanzees and bonobos [Dainton and Macho 1999]).

Locomotion in Basal Hominins and Early Human Ancestors

If the LCA was a knuckle walker, were basal hominins also knuckle walkers? There is virtually no information on the locomotor anatomy of basal hominins. Evidence in favor of the idea that basal hominins were knuckle walkers includes the fact that all the African apes are knuckle walkers; both the *Dryopithecine* fossil apes and the earliest human ancestors display features in the arm, elbow, wrist, and hands that are unique to knuckle walking (Richmond 2000; Richmond, Begun, et al. 2001).

"African apes and hominins share many forelimb characteristics that may be functionally related to knuckle-walking, and yet are retained in humans because they are compatible with bipedalism and the functions required of the hominin upper limbs" (Richmond, Begun, et al. 2001, 83). Most of these features are related to each other, and involve two functions: (1) stabilizing the wrist in an extended position, and (2) reducing stress from compression of the knuckles. These are achieved partly through enlarged joint surfaces, reduced movement through bone fusion, and strongly developed wrist flexors.

Gordon Hewes's Food Transport Scenario for Bipedalism

Gordon Hewes (1961, 687) begins by noting that "upright posture and bipedal gait have been considered outstanding features of human nature" since the time of Haeckel and Darwin. After reviewing various discussions of bipedalism, he says that none of the authors he cites has explained why bipedalism arose. He notes that work on the fossil australopithecines demonstrates that bipedalism preceded brain enlargement, contrary to earlier scenarios that saw cerebral expansion as the prime mover of hominization. He argues that whatever its cause, bipedalism must have involved "rapid and ruthless" selection judging from the back problems that plague humans today.

His answer to the question of why certain primates already capable of sporadic bipedalism became habitually bipedal is "effective use of a new food resource required its transport over considerable distances, and only bipedal locomotion, *by freeing the arms and hands for carrying*, could achieve maximal transportational efficiency" (Hewes 1961, 689). He dismisses the idea of sudden changes from a mutation, saying that models should explain the advantages of incremental changes in locomotion. Consistent with this, he argues that early bipedalism must have coexisted with the continuing ability to climb trees as suggested by the more flexible, long-heeled foot of *Australopithecus*.

Disagreeing with Darwin (1891) and Washburn (1960) and others that bipedal locomotion evolved for tool use, he argues that 80 percent of tool use is done in a sitting or squatting or kneeling position and therefore does not require bipedalism. He also disagrees with Dart's argument that bipedalism evolved from bipedal throwing and clubbing because, based on the fossil and archaeological record, Hewes doubts that early hominids (now called hominins) were skilled hunters or tool users. He also dismisses the visual surveying and the upright-display explanations on the grounds that these

---

Related features occur in the arm bones of early human ancestors *Australopithecus anamensis* and *A. afarensis* and *Ardipithecus ramidus* (Richmond, Begun, et al. 2001). Whether they actually engaged in knuckle walking is uncertain. Unfortunately, relevant fossil remains of forelimbs are lacking for the earliest basal Hominini, *Sahelanthropus tchadensis* and *Orrorin tugenensis*, so it is unknown whether they displayed these features as well. Evidence for retention of knuckle-walking anatomy in basal hominins is consistent with other morphological and molecular evidence that chimpanzees and bonobos are our closest living relatives (Richmond and Strait 2000). Moreover, the

activities would occur sporadically at best. Citing the common explanation of carrying tools or infants, Hewes says, "My disagreement is only over the nature of the burdens carried by the incipiently bipedal protohominids and the relations of this carrying to the emergence of habitual bipedalism" (1961, 693). He concludes that "the only activity likely to have had the capacity to transform a mainly quadrupedal ground-dwelling Primate into an habitual biped would have been food transport from the places where food was obtained to a home base where it was consumed" (698).

Regarding the environment, Hewes (1961, 700) says, "Contemporary paleoanthropological opinion places the transition to hominid status in tropical park-savanna lands, where narrow forest environments extend along river-courses, flanked in grassy planes." Following Bartholomew and Birdsell (1953) and others, he suggests that protohominids were scavengers, that is, "meat and marrow eaters but seldom actual hunters" (695). Following Spuhler, he argues that the adoption of a carnivorous diet with its concentrated food sources invited transport. Finally, he cites Oakley in arguing that drought in East Africa could produce abundant food in the form of half-eaten carcasses of large ungulates: "The first primates to become accustomed to this new kind of food would have embarked on a momentous journey" (201).

Dismissing tool use as involving little intelligence, Hewes places the rise of habitual tool use after bipedalism: "Out of the habit of carrying carrion to a lair could have arisen the first regular rather than sporadic use of tools" (1961, 703). He suggests that scavenging would have laid the groundwork for a sexual division of labor occurring under the adoption of hunting. In the course of his argument, he also mentions experiments he is performing on conditions favoring greater use of bipedalism in captive macaques.

---

closer fossil hominins are to the LCA, the greater the likelihood that they retain ancestral characteristics.

Locomotion of Early Human Ancestors: Upright Standing or Walking? Throwing or Carrying?

How did australopithecines stand and move? Early anatomical studies emphasized their bipedality, which was confirmed by fossil footprints at Lateoli in East Africa (M. Leakey and Hay 1979; Tuttle, Webb, et al. 1991). Later studies of additional fossil material suggest that their bipedalism differed significantly from that of modern humans. These studies have revealed that australopithecines display a mixture of derived bipedal anatomy in the lower limbs combined with retention of some primitive characteristics in the hands and feet that suggest continued capacity for tree climbing. These include shorter legs relative to longer arms, flatter, wider pelvises (McHenry 1986; McHenry and Berger 1998), and more upward-facing shoulder joints than later hominins. They also display longer, more curved fingers and toes than those of middle-period hominins (Susman, Stern, et al. 1984), and inner-ear canal orientation characteristic of quadrupeds rather than bipeds (Spoor, Wood, et al. 1994). Some anthropologists have suggested that shorter legs, longer arms, upward-facing shoulder joints, and longer, more flexible feet were used in tree climbing (Susman, Stern, et al. 1984).

The recently discovered 3.3-million-year-old baby *Australopithecus afarensis* from Dikika, Ethiopia, displays these mosaic features of bipedal feet and legs combined with long arms, curved fingers, and a shoulder blade and upper arm morphology reminiscent of gorillas suggesting "arboreal behavior" (Zeresenay, Spoor, et al. 2006). Could this have included brachiation as well as climbing? This interpretation is consistent with Rose's (1991) point that we should expect persistence of earlier patterns in the early stages of evolution of bipedalism. It is also compatible with paleoecological reconstructions indicating that these creatures lived in woodland habitats. See figure 5.3 for a depiction of *A. afarensis* anatomy compared to chimpanzee and human anatomy.

Locomotion in Middle-Period Human Ancestors

Clearly bipedalism changed from one of several modes of locomotion, including tree climbing among australopithecines, to become a singular mode of habitual bipedalism among *H. ergaster/erectus* that lived from about 1.8 million years ago and moved out of Africa into Asia.

Analyses of the skeleton of the Nariokatome *H. ergaster* or Turkana Boy have revealed a major change in these creatures in the direction of complete commitment to bipedal locomotion (Walker and Leakey 1993; Walker and Shipman 1996). This largely complete skeleton of a young boy has been invaluable in revealing the anatomy of bipedal locomotion in this species. First of all, it revealed a modern height combined with modern proportion of arm to leg length rather than the diminutive stature and longer arms than legs typical of earlier forms. Second, it revealed large robust femurs with a modern angle and knee. Third, it revealed a broad and shallow chest morphology rather than the narrow and deep chest inverted-cone morphology of the chests of apes and earlier hominins. Fourth, it revealed a more modern pelvic

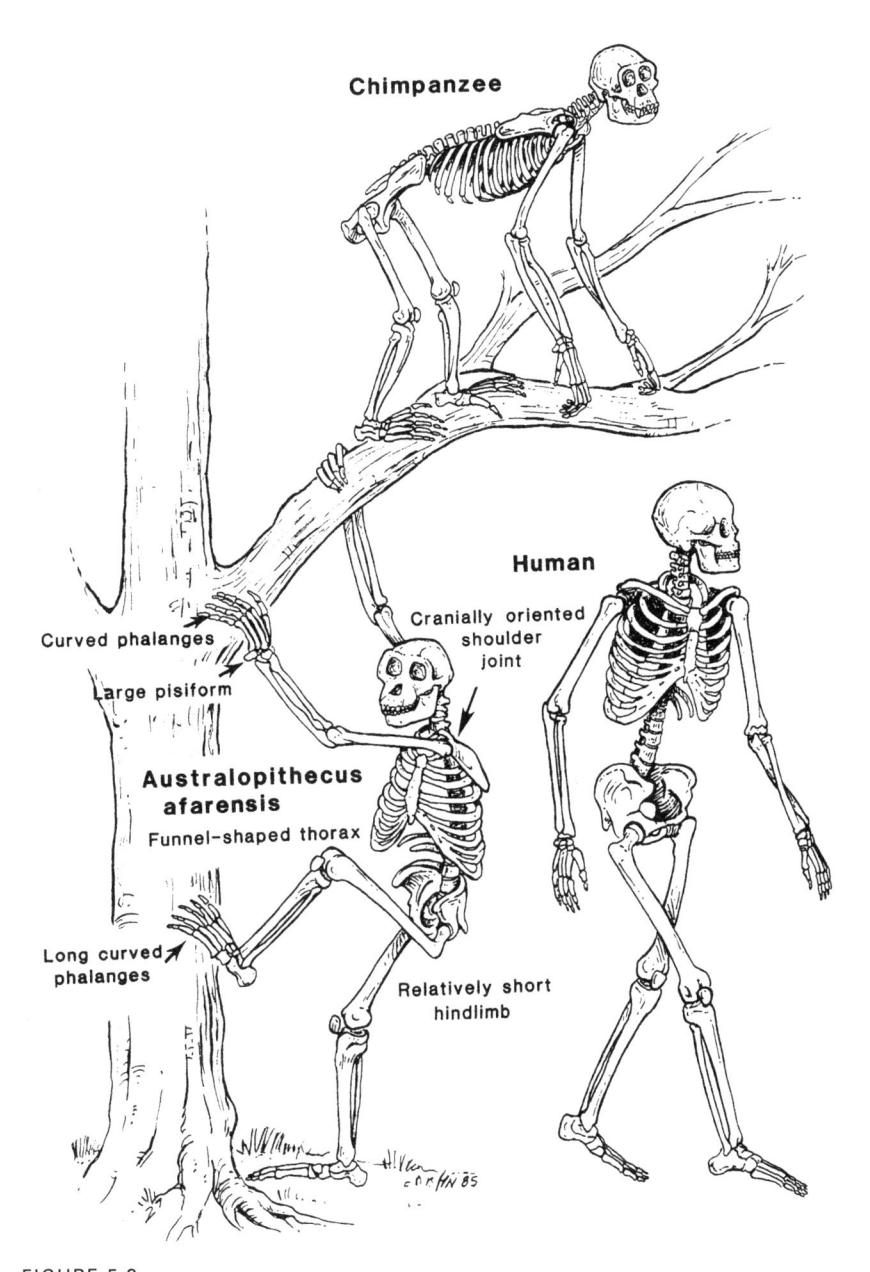

FIGURE 5.3.
Comparison of *Australopithecus afarensis* anatomy with chimpanzee and modern human anatomy.
Source: Drawing originally published in Fleagle, J. G. 1999. *Primate adaptation and evolution.* New York: Academic Press, p. 516. Copyright © Academic Press 1999. Reprinted with permission.

shape and size compatible with the larger brain size of these creatures rather than the smaller flattened pelvises of australopithecines. Fifth, it revealed shortened fingers and toes specialized for walking rather than climbing. Despite his larger brain and body size, the life history of this boy was still more like those of australopithecines than those of modern humans (C. Dean, Leakey, et al. 2001; see chapter 6 on life histories).

Based on this and other evidence, anthropologists agree that these and other middle-period human ancestors were efficient, striding, long-distance walkers and runners. They also agree that they manufactured Acheulean tools and were scavengers and hunters of large game and perhaps excavated roots and tubers (see discussion in chapter 4). They lived in savanna as well as other habitats that exposed them to predators. Presumably their larger body size and more sophisticated tool use as well as use of fire allowed them to cope with this challenge without resorting to tree climbing. These characteristics are compatible with Wheeler's (1984) idea that upright posture protected the body from overheating in the midday sun by minimizing the surface area exposed to solar radiation, but not with his proposal that these pressures led to the evolution of bipedalism (see chapter 7).

SCENARIOS FOR THE ORIGINS OF BIPEDALISM

Owing to the very recent discoveries of basal hominins dated before three million years ago, most of the scenarios for the origins of bipedality focus on the appearance of this characteristic in early human ancestors living between three and one million years ago. Charles Darwin (1871) suggested that bipedalism arose to free the hands for making and using tools in offense and defense against predators and prey and competitors. Friedrich Engels (1896) concurred with Darwin's scenario, emphasizing that the hands are both the organ of labor and its product.

Like Darwin, many later scenario builders envisioned a complex of features that evolved in relation to bipedalism. Like Darwin's, this complex often included bipedalism, reduced canine teeth, and tool use. In many cases these scenarios were based on the hypothesis that early human ancestors made their living by hunting and/or scavenging as well as gathering. Although several other dietary hypotheses have also been suggested to explain bipedalism, the hunting hypothesis has been one of the most persistent and also the most contested vision of early hominin subsistence.

Raymond Dart (1953, 1959) concurred with Darwin's scenario. He emphasized the importance of bipedal stance, hand, and brain specialization for stone throwing and club swinging in hunting and fighting. He bolstered his argument by comparing the accuracy of human throwing with the inaccuracy of great ape throwing. Kortlandt and Kooij (1963) also compared throwing skills in apes and humans, noting the fair accuracy of ape throwing. Like Darwin, they argued that early humans engaged in aimed throwing to defend themselves against predators. Both Mary Marzke and Richard Young have elaborated this scenario (Marzke 1986; Marzke, Longhill, et al. 1988; Young 2003). Paul Bingham (1999) argues that aimed throwing and clubbing are key hominin adaptations for killing at a distance, leading to punishment of cheaters in nonkin groups (see chapter 10).

Based on their movement studies, Marzke, Longhill, et al. (1988) argue that the large buttocks muscle, gluteus maximus, characteristic of humans, is only minimally involved in bipedal walking; it is strongly involved in a fixed stance on the hind limbs and rotation and braking of the trunk during throwing, clubbing, digging, and lifting by the forelimbs (all involving major shifts in the center of gravity). Bingham (1999) cites Marzke's work on the gluteus maximus in his argument for the evolution of aimed throwing and clubbing.

Roger Westcott's Upright Display Scenario

Roger Westcott (1967) suggests that hominids (now called hominins) differed from pongids primarily in being bipedal and "preferentially carnivorous" rather than quadrupedal and herbivorous. Noting reports of bipedal threat displays by apes in trees and on the ground, he says that the "apparent purpose of bipedal behavior is to make the bipedalist appear taller, and therefore to intimidate potential predators or antagonists" (738). He follows L. S. B. Leakey in assuming that these early "Kenyapithecus type" hominids diverged from apes about twenty million years ago. He argues that their chief predators and competitors included oreopithecid apes, pongids, bears, hyena-like creodonts, and big cats. He proposes that, like pongids, hominids therefore had agonistic interactions that included "two-legged standing or running, probably accompanied by fist-shaking or arm-waving, and possibly involving the seizing and brandishing of sticks or stones" (738). He also argues that these activities would have been "both frequent and impressive" (738). He notes that bipedal displays probably were only one cause of bipedalism, other causes including carrying tools and peering over tall grass while scavenging carnivore kills.

Earlier studies by Marzke revealed two grips used in throwing, digging, and clubbing: the pad-to-side pinch of a stone between the thumb and side of the index finger and the three-jaw chuck grip involving the thumb, index finger, and the third finger. She notes that "there are two sets of derived features in the modern human hand which seem to find an explanation in requirements for control of these grips and the forces endured by the palm of the hand during these activities" (Marzke 1986, 204–205).

She adds that features involved in the three-jaw chuck grip used in throwing were already present in *A. afarensis*. Likewise, these creatures were capable of pad-to-side pinch despite their relatively shorter thumbs: "The ability to grip firmly and control the position of a small stone by the thumb, index and middle fingers may have allowed them to throw the stone with sufficient aim and velocity to stun or kill small animals" (Marzke 1986, 208). She also notes that these changes in the hands preceded the appearance of worked stone tools, though see discussion of possible association of *Paranthropus*'s hand with tools (Susman 1988, 1998). In any case, it is telling that throwing is more strongly right-handed than other forms of tool use (Calvin 1983).

Young elaborates on this, arguing that both sexual and natural selection favored reproductive success in males proficient in aimed throwing and clubbing in scavenging, hunting, and male combat: "Selection for improved throwing and clubbing produced an innovative, instinctive, whole-body motion performed from an upright stance that begins with a thrust of the legs. Improved dynamic upright balance on more powerful legs and resilient feet in the service of throwing and clubbing would have made upright locomotion more efficient, leading to its increased use and eventually culminating in habitual bipedalism" (Young 2003, 166).

Aimed throwing and balance were also associated with specialization of the cerebellum and the cortex for precise timing of throwing motions, which may have been an exaptation for language (Calvin 1983). This scenario raises questions about the source of appropriate stones for use as aimed missiles. Perhaps it may explain the occurrence of stone caches as proposed by Potts (1984), and the occurrence of round bola-like stones at Olduvai Gorge, Tanzania.

In contrast to those emphasizing standing, Gordon Hewes (1961, 1964) focused on bipedal walking, proposing that this behavior was favored for carrying a new source of food, probably "carrion," that is, scavenged meat. He disagreed with Dart's view, arguing that tool use and hunting were poorly developed in early hominids (known as hominins in modern taxonomy). Others who favored the carrying hypothesis suggest a variety of items that might have been carried, ranging from tools or weapons to hunted or scavenged meat (Dart 1959; Hewes 1964; Washburn and Lancaster 1968; Lancaster 1978), to gathered and/or extracted foods (Bartholomew and Birdsell 1953; N. Tanner and Zihlman 1976; Parker and Gibson 1979; Zihlman and Tanner 1979), to dependent infants (Etkin 1954). In other words, the carrying hypothesis has been associated with tool use in both hunting/scavenging and gathering.

Sally Linton (1971) and others (N. Tanner and Zihlman 1976; Zihlman and Tanner 1979; Lovejoy 1981) argued that bipedalism was for carrying gathered foods in containers. Bartholomew and Birdsell (1953) and Dart (1953) suggested that human ancestors evolved an upright stance for picking fruit from low trees or for digging for roots and tubers as well as for scavenging. The fruit-picking idea was later championed by Hunt (1994) and the root and tuber excavation by O'Connell (O'Connell, Hawkes, et al. 1988) and Wrangham, Jones, et al. (1999). Alister Hardy (1960) suggested that bipedalism evolved for gathering foods from the sea and seashore, an idea later championed by Elaine Morgan (1982) (see chapter 7).

In contrast to ideas about the dietary and tool-using complex, is Roger Westcott's (1967) view that bipedalism originated in upright threat displays favored by sexual selection. This idea was recently elaborated by Nina Jablonski and George Chaplin (1993). Another sexual selection scenario proposed that hominid males competed by presenting the brains of prey as nuptial gifts to females (Parker 1987).

Evaluating Alternative Hypotheses

Most of these scenarios are plausible in that they are extensions of behaviors in our closest living relatives, African apes. This satisfies the continuity principle. The continuing existence of bipedalism in several related species satisfies the phyletic criterion for adaptation (bipedalism is a shared derived feature among hominins; see chapter 2). But what is the nature of the adaptation? Which selection pressures favored bipedalism? This is a difficult problem because all of the proposed functions—throwing, carrying, gathering, upright displays—are plausible. Indeed, they are all observable today. The real question is, which function was earliest and primary?

Richard Young's Throwing and Clubbing Scenario

Richard Young (2003) begins with the assertion that although "there is general agreement" that the evolution of the human hand is linked with tool behavior and bipedal gait, there is no agreement on what kinds of tools were used for what purposes. He proposes that "the tools were hand-held weapons that were hurled or swung as bludgeons at adversaries during disputes, providing the aggressors with advantages that in various ways promoted reproductive success" (165). He argues that the best throwers would rise in dominance, that they would prevail in territorial disputes and in contests with predators while scavenging. All these successes would give them reproductive advantages in mating. He focuses on two concomitants of throwing and clubbing: the evolution of bipedalism and of the power and precision grips of the hand.

First, he argues that "selection for improved throwing and clubbing produced an innovative, instinctive, whole-body motion performed from an upright stance that begins with a thrust of the legs, improved dynamic upright balance on more powerful legs and resilient feet in service of throwing and clubbing would have made upright locomotion more efficient, leading to increasing use and eventually culminating in habitual bipedalism" (Young 2003, 166). He also notes that "the throwing and clubbing motion that begins in the legs progresses through the hips, torso, and arm, and ultimately imparts accumulated kinetic energy to the hand or hands holding the weapon" (166).

Second, he argues that the two uniquely human grips, the power and the precision grips, arose as adaptations for clubbing and throwing, respectively. The human hand contrasts with that of the chimpanzee in

---

As noted above, many authors from Darwin on have mentioned throwing and or clubbing in lists of activities associated with bipedalism. No one, however, has focused specifically on these actions as Marzke, Bingham, and Young have. No one focused on stance as opposed to locomotion as they have. We believe that this focus on *stance* as opposed to locomotion reveals that the major selection pressure was aimed throwing and clubbing.

We favor the throwing and clubbing scenario as the primary factor in the evolution of bipedalism for several reasons. First, aimed throwing and clubbing are new behaviors derived from aggressive bipedal displays rather than intensification of such common chimpanzee-like behaviors as object carrying

having a longer, larger thumb that is not only more mobile and muscular, but also fully opposable. Its fingers contrast with the long curved fingers of chimpanzees in their shortness and straightness as well as their graduated length. Human fingertips have large pads and they are fully opposable with each other and the thumb and rotate inward toward the thumb when they are flexed. Three new muscles aid in this. The palm has also been strengthened. These derived features combine to facilitate the power and precision grips. Young (2003, 169) argues that "for efficient throwing the hand must be able to grip the missile while energy is transmitted to it, then accurately control its release. The thumb must be long enough and sufficiently mobile to oppose its fingertip pads to the missile on one side while the fingers oppose their distal pads to the opposite side and adjust themselves to irregularities in naturally occurring rock spheroids." He further notes that letting go of the missile within the millisecond that the arc of the moving hand is in the position that allows it to hit its target, requires precise control by the hand and the brain. Likewise, he argues that powerful clubbing is facilitated by gripping obliquely along a cylindrical cavity formed diagonally across the palm (bounded by a fat pad and palmar muscle on the ulnar side, and another muscle and fat pad on the radial side) and by the twist of the wrist, which aligns the club handle with the forearm. Both actions are served by the finger rotation, thumb length, and distensible finger pads. Young argues that the hand of *Australopithecus afarensis* had changes from chimpanzee hands that rendered them effective for throwing and clubbing grips. Since no stone tools have been associated with these fossils, he concludes that throwing and clubbing antedated stone tool production.

and use of tools as hammers and probes. Second, these behaviors selected for strengthened trunk and arm muscles and for improved balance, which would have made habitual bipedal locomotion practical. Third, these behaviors could explain the concurrent evolution of the gripping hand and handedness, which cannot be explained by stone tool production. Fourth, these anticompetitor and antipredator defense behaviors could explain canine reduction already apparent in these small creatures. Fifth, aimed throwing in male competition explains the evolution of the enlarged chest and upper arm muscles, which are derived secondary sexual characteristics in human males. Hence, it converges in outcome with possible sexual selection for male upright threat displays (Westcott 1967; Parker 1987; Jablonski and Chaplin 1993). It is also consistent with a scenario for killing at a distance with aimed missiles (see chapter 10).

Finally, skills involved in throwing and clubbing would have enhanced other forms of tool use such as digging for deep roots and tubers and other forms of extractive foraging, as well as carrying, which must have elaborated along with scavenging and small game hunting.

CONCLUSIONS ON THE EVOLUTION OF BIPEDALISM

Recent analysis of early hominin fossils suggests that they had a knuckle-walking ancestor, derived from their last common ancestor with African apes. The foregoing discussion suggests that the evolution of bipedalism occurred in at least two major steps: The first step was a primitive form of terrestrial bipedal locomotion combined with preexisting arboreal locomotion (climbing and perhaps suspensory movements) seen in early human ancestors, the australopithecines; the second step was a committed form of terrestrial striding bipedal locomotion, first seen in middle-period human ancestors, *H. erectus*. This purported sequence satisfies the continuity criterion.

We favor the scenario, originally suggested by Darwin, that bipedalism was initially favored as part of an adaptation for specialization of the trunk and hand for aimed throwing and clubbing, providing an alternative to offense and defense with elongated canine teeth. Bipedal stance probably preceded bipedal locomotion. Aimed throwing and clubbing undoubtedly co-occurred with continuing use of tools for a variety of other tasks including extractive foraging and carrying. The second stage of evolution of committed bipedal locomotion apparently arose as an adaptation for long-distance walking and running associated with scavenging and hunting with worked stone tools.

As has been realized since the recognition of the Taung baby as an early human (Dart 1925), the evolution of bipedalism preceded the evolution of the enlarged brain. This is one of many cases of so-called mosaic evolution in which different functional systems evolved at different rates during human evolution.

6

Origins of Human Life History

Each organism reaches maturity through a longer or shorter course of growth and development: the former term being confined to mere increase in size, and development to changed structure. The changes may be small and insensibly slow, as when a child grows into a man, or many, abrupt, and slight, as in the metamorphoses of certain ephemerous insects.

—*Darwin 1871, 383*

As discussed in earlier chapters of this book, paleoanthropologists have long recognized the importance of locomotor behaviors (chapter 5) and subsistence strategies (chapter 4) in human evolution. This is not surprising, considering the hard evidence, for example, footprints, fossil legs, feet, pelvic girdles, spines, and stone tools associated with fossil animal and plant remains. Life cycle or life history, an equally important aspect, was often overlooked or misunderstood because it is difficult to infer from the fossil record (see also Lovejoy 1981). New methods allow researchers to infer hominin growth patterns and the duration of developmental periods from fossil teeth.

Study of four fossilized skeletons of immature hominins, the Taung child (*Australopithecus africanus*), the Turkana boy (*Homo ergaster*, the early African form of *H. erectus*), the Dederiyah youngsters (*Homo neanderthalensis*), and the newly discovered Dikika baby (*A. afarensis*), has been especially critical to understanding hominin life histories. These and other partially complete skeletons such as Lucy (*A. afarensis*) also reveal body size and sexual dimorphism. These studies have revolutionized the picture of human evolution, casting doubt on the idea that modern human life history developed early in hominin evolution. These new methods have revealed evidence for shorter, more apelike developmental periods in *Australopithecus*, early *Homo*, and even *H. erectus*.

Life History Configurations

The study of life history focuses on age-specific mortality and fecundity in relation to features of the life cycle, including how long individuals of a species grow, how large they get, when they begin reproducing, how many offspring females can produce in a lifetime, and how long they live (see also B. Smith and Tompkins 1995). Each species has a unique combination of life-cycle variables (Millar 1977; Millar and Zammuto 1983; Harvey and Clutton-Brock 1985; Harvey, Martin, et al. 1987).

Because energy is limited, it must be allocated among all aspects of life, including growth and development, maintenance, defense, and reproduction. Therefore, the life history of a given species reflects a number of biological trade-offs (see also Harvey and Read 1988; B. Smith and Tompkins 1995; Leigh and Blomquist 2007). For example, a species can produce relatively few offspring and invest heavily in each offspring (K configuration), or produce many offspring and invest relatively little in each one (r configuration; see table 6.1). It is impossible to produce large numbers of offspring and invest heavily in each one because there is limited energy to devote to reproduction (i.e., mating, gestation, and rearing). Producing a large number of offspring leaves relatively little energy for offspring care.

The K configuration is usually associated with single well-developed (precocial) neonates, born with eyes open, able to walk or cling, and well insulated (Portmann 1990). In contrast, the r configuration is associated with large lit-

Table 6.1. Variables Defining r-Configured versus K-Configured Species

Variable	r-Configured	K-Configured
Time of first reproduction	Early	Late
Litter size	Large	Small
Time of maturity	Early	Late
Life span	Short	Long
Parental investment	Low	High
Infant mortality	High	Low

ters of so-called fetal (altricial) infants born with eyes and ears closed, hairless, and unable to walk. Rodents and carnivores and other nesting or burrowing species display the fetal pattern, whereas most free-ranging ungulates and primates display the precocial pattern. This framework is most useful when comparing species and populations within the same order (e.g., the order Primates, the order Carnivora) rather than distantly related species.

Most monkeys and apes show the K configuration and give birth to single precocial young. Adolph Schultz (1969), a pioneer in the study of primate life histories, described the stepwise increase in the duration of life history stages from prosimians to monkeys to apes to humans (see figure 6.1; table 6.2).

Some recent work casts doubt on this simple picture of the evolution of stepwise increases in stages of primate life histories and tightly correlated features of life history. Some developmental biologists suggest that evolution acts on dissociable modules. In other words, various elements can change at different rates, that is, they are subject to mosaic evolution (Leigh and Blomquist 2007). This seems to be true of humans as compared to other primates.

Even among primates, humans fall at the extreme end of the life history continuum in most features (table 6.2; see also Hawkes, O'Connell, et al. 1998), displaying longer maximum life spans and later ages at puberty and first reproduction than our closest living relatives, the great apes (Sacher and Staffeldt 1974; B. Smith and Tompkins 1995). Humans also show other deviations from typical primate life history features in gestation, infancy, juvenility, and reproductive maturity (see also B. Smith and Tompkins 1995).

Compared to other primates (except great apes), humans display three additional stages, *childhood, adolescence,* and *menopause* (Bogin 1990; Bogin 1999); both adolescence and menopause are nonreproductive periods. In addition, the human gestation period is significantly shorter than would be expected from brain size and life span. Consequently, human infants have been called secondarily altricial. Likewise, under favorable conditions, the birth interval is shorter. Finally, human males invest more in parental care than males of other primate species. (Parenthetically, it is important to note that life history patterns in males in polygynous species differ from those of females, reflecting their differing mating strategies. In addition to their longer immaturity, they are significantly larger than females. Males in these species typically invest all of their reproductive energy in mating and little or none in parenting [Trivers 1972]). See also chapter 7.

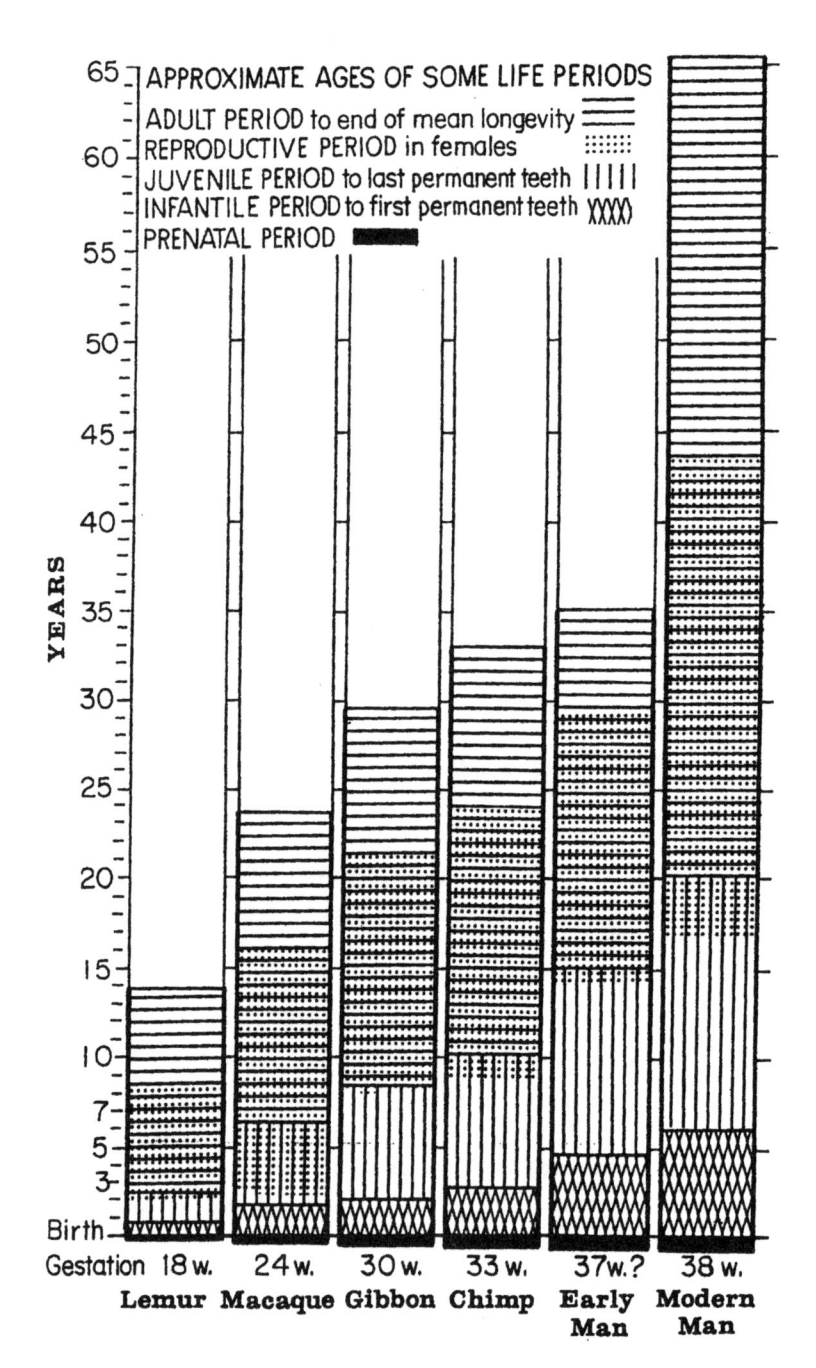

FIGURE 6.1.
Comparison and contrast of life stages in various primate species.
Source: B. H. Smith. 1992. Life history and the evolution of human maturation. *Evolutionary Anthropology* 11 (4): 136. Copyright © 1992 John Wiley & Sons, Inc. Reprinted with the permission of Wiley-Liss, Inc., a subsidiary of John Wiley & Sons, Inc.

Table 6.2. Comparison of Common Life History Variables among Some Primates

Variable (in years, unless otherwise noted)	Lowland Gorilla	Chimpanzee	Human
Age at puberty (M; F)	10; 6.5	13; 11.25	12–16; 10–15
Age at first birth	9.3	13	16–20+
Interbirth interval	4	5	0.8–4
Age at weaning	3.5	5	2
Average body weight (lbs) (M; F)	370; 190	110; 90	150; 120
Adult brain weight (cc)	506	410	1,250
Maximum life span	50	53	100

Sources: Jungers 1985; Harvey, Martin, et al. 1987; Harrison, Tanner, et al. 1988; Malina and Bouchard 1991; Ross 1991; Watts and Pusey 1993.

Later in this chapter we discuss three hypotheses that have been proposed to explain human life history, the pair bonding, parental investment, and grandmothering scenarios. Before addressing these, we review techniques for assessing the ages at death of hominin fossils, and thereby estimating their life histories.

Methods Used to Predict Age at Death in Fossil Hominins

How did the life histories of our hominin ancestors compare with those of modern humans, and when did the modern human life history pattern appear? In order to answer these questions, paleoanthropologists need to reconstruct the ages at death of immature hominin fossils. Anthropologists have used two primary techniques for this task: dental and bone analysis.

Tooth Formation and Long Bone Growth Patterns

Lake other mammals, primates have two sets of teeth: deciduous or baby teeth, and permanent or adult teeth. Adult Old World monkeys and apes share the same derived dental formula of thirty-two teeth (counting each quarter of the upper and lower jaws: two incisors, one canine, two premolars, and three molars). They have twenty deciduous or baby teeth (counting five in each quarter of the two jaws: two incisors, one canine, two baby molars). Premolars are actually numbered 3 and 4 rather than 1 and 2 to reflect which of the four premolars of ancient mammalian ancestors were retained in these primates (Clark 1959).

Lovejoy's Scenario for Origins of Human Life History, Subsistence, and Bipedalism

Owen Lovejoy (1981) proposes to explain the origins of bipedalism, pair bonding, and human life history as adaptations for provisioning females and offspring by monogamous male gatherers.

He begins his scenario by noting that whereas models of human origins "must directly address the few primary differences separating humans from apes," they must also address when these characters first appeared, as revealed by the fossil record (Lovejoy 1981, 341). He notes that canine reduction and bipedalism were already present in *Australopithecus afarensis* more than three million years ago, and worked stone tools first appear about a million years later at two million years ago. Based on this, he argues that "it is likely that the earliest hominids made no use of tools at all, or that such use was comparable to that in other extant hominoids and was not critical to their survival or pivotal to their evolution" (341). He also notes that brain expansion first occurred between two million and three million years ago. He argues that although *A. afarensis* probably walked like modern humans, their pelvises display no evidence of accommodation to large fetal brains. In summary, he says, "Only bipedal locomotion and partial dental modifications can be shown to have an antiquity even approximating the earliest appearance of unquestioned, developed hominids (*A. afarensis*)" (343).

Turning to hominid (now called hominin) paleoecology, Lovejoy notes that present evidence suggests that the hominid clade in the Miocene evolved

---

There are two primary components of tooth development: formation (of crowns and roots) and eruption through the gums (B. Smith 1991b). Completion of adult dentition is correlated with the closure of the growth cap on long bones in a wide range of mammals (Shingehara 1980). Therefore, it can serve as an indicator of the beginning of adulthood (B. Smith 1989). Studies of the sequence of dental eruption in various primates, including the great apes and humans, show that the age of eruption of the first, second, and third molars (M_1, M_2, and M_3), follows a specific pattern shown in table 6.3 (Hurme 1949; Nissen and Riesen 1964; Moorrees and Kent 1978; Willoughby 1978).

Because of the strong association between dental eruption and skeletal growth (Alvarez and Navia 1989), dental eruption is useful for determining the age of mammals. Dental formation appears to be the more robust measure than dental eruption because external forces such as nutritional stress and

in forest or mosaic habitats rather than in grasslands and savanna. After criticizing earlier hominid scenarios, he summarizes some demographic principles as a prelude to introducing his own "behavioral model for early hominid evolution" (1981, 344). First, he suggests that new adaptations must have evolved from preadaptations of Miocene apes. Second, he suggests that Miocene hominids were omnivores, who when faced with variable mosaic environments would have increased their foraging ranges. This increased foraging distance, combined with pressures on bipedal females to carry their "altricial" offspring, would have favored at least partial separation of male and female day ranges. Males would benefit from increased parental investment, "collecting available food item or items and returning them to the mate and offspring" thus favoring monogamous pair bonding (1981, 345). Lovejoy argues that this would increase infant survivorship and reduce the birth interval.

In his model for the origins of bipedalism, he argues that carrying significant amounts of food would have favored "consistent, extended periods of upright walking" (Lovejoy 1981, 345). He also suggests that this behavior would have favored the development of material culture, particularly carrying devices. These adaptations would have helped early hominids meet the "demographic dilemma" of increasing investment in offspring lowering reproductive rates (346). He also argues that this model explains the following "pair bond enhancers": the loss of estrus, the unusual pattern of sexual dimorphism in humans, which involves larger male body size combined with canine reduction, and epigamic features in females.

body size can affect tooth eruption (Alvarez, Lewis, et al. 1988; Alvarez and Navia 1989). Moreover, tooth formation is highly heritable (Lewis and Garn 1960; Garn, Lewis, et al. 1965b; Moorrees and Kent 1978), and so expresses lower degrees of variation than skeletal development (Lewis and Garn 1960). Tooth formation also appears to be resistant to nutritional changes (Lewis and Garn 1960; Prahl-Andersen and Van der Linden 1972) and increases in height

Table 6.3. The Age (in years) of Dental Eruption in Lowland Gorillas, Chimpanzees, and Humans

Dental Eruption	Lowland Gorilla	Chimpanzee	Human
First molar (upper/lower)	3.5/3.5	2.9/2.9	6.3/6
Second molar (upper/lower)	6.5/6.6	6.2/6.3	12/11.5
Third molar (upper/lower)	11.4/10.4	10.3/9.3	20.5/18

Sources: Dahlberg and Menegaz-Bock 1958; Willoughby 1978; Conroy and Mahoney 1991; Schwartz, Mahoney, et al. 2005.

(Helm 1969). The fact that dental formation is relatively unaffected by these factors makes it an excellent measure for estimating ages of our extinct ancestors.

The growth of tooth enamel and dentin (the substance below the enamel) is regular enough to allow fairly accurate estimates of time elapsed during formation of tooth crowns, hence the age at death (D. Fisher 1987; Hohn, Scott, et al. 1989). The main problem with this method is that it requires cutting teeth into sections to count the growth bands (e.g., Hohn, Scott, et al. 1989). Counting stripes formed within the occlusal and lateral enamel every nine days (on average) gives an estimate of time of crown formation (C. Dean, Leakey, et al. 2001). Bromage and Dean (1985) have found an alternative to this invasive and destructive method of estimating time elapsed during crown formation. They found that growth bands visible on the surface of well-preserved human teeth could also be seen in the teeth of fossil hominins (see also C. Dean 1985; M. Dean 1987). This method yields the most direct and accurate estimate of age at death.

Tooth formation sequences also can yield an estimate of an individual's *dental age* at death (B. Smith 1991b). This is useful for determining growth and development of both extinct and modern organisms for several reasons. First, age estimates based on tooth formation are reliable (Garn, Lewis, et al. 1965a; Garn, Lewis, et al. 1965b; Niswander and Sujaku 1965) because dental development is little affected by endocrine disorders and other developmental problems (B. Smith 1991b). Second, dentition is closely incorporated into the overall pattern of growth and development because teeth must erupt in a specific order for proper functioning (B. Smith 1991a). Third, since dental development correlates with life-cycle stages of mammals, it provides an important clue to the growth of extinct species. Overall, it is crucial for determining developmental stages of extinct hominins because the first molar (M_1) erupts at age of weaning, the second molar (M_2) erupts at puberty, and the third molar (M_3) at the onset of reproduction (B. Smith 1991b).

The sequence of dental formation (of crowns and roots) can be obtained from X-rays of the upper (maxilla) and lower (mandible) jaws, or by looking at the roots and crowns of exposed teeth in broken jaws of juvenile fossils (B. Smith 1991b). In this method, individual teeth are first scored according to their developmental stage and then arranged according to their sequence of development. Finally, they are assigned an estimated dental age using averages based on human or great ape growth standards, and then plotted on developmental charts for each species (B. Smith, Crummett, et al. 1994). The degree of development for each tooth is compared to the reference population growth standard. Thus, investigators can assess the similarity of the sequence of dental development of individuals compared with those of living "reference populations."

If the dental development pattern for an individual fits the reference population growth standard, this indicates that the degree of variation of dental development between the two is relatively low. If the pattern does not fit the reference population, the degree of variation of dental development is relatively high, meaning that the individual displays a different dental development pattern than the reference population (Smith 1991a).

Unfortunately, however, the sequence of dental development is not equivalent to the timescale of dental development, which depends upon the underlying growth processes of enamel. African apes and early hominins have similar enamel growth trajectories (even though African apes have thin enamel and early hominins have thick enamel). As indicated above, when direct counting of enamel growth lines is impossible, estimates of developmental ages of fossils depend upon which reference species is selected. Although *H. erectus* (and *H. ergaster*) has an identical sequence to modern humans, their dental development is more similar to that of African apes and early human ancestors. The identical sequence does, however, suggest that the shift in timing affected all the teeth equally (C. Dean, Leakey, et al. 2001).

Age at death of immature fossils can also be determined using growth charts for the timing of closure of the growth plate on long bones (B. Smith 1993; Byers 2005). As with dental growth charts, both modern human and chimpanzee charts can be used for reference. This technique is less accurate than the dental techniques, but provides complementary results. Even partially complete adult skeletons can yield important information regarding body proportions, body size, and sexual dimorphism. As we reported in chapter 5, australopithecines had long arms relative to legs. *A. afarensis* was also highly sexually dimorphic, males being 40 percent larger than females. *Homo erectus* males were thought to be 20 percent larger than females (McHenry 1992; McHenry 1996). The recent discovery of a small *H. erectus* skull in Kenya, however, suggests that they may have been highly sexually dimorphic (Spoor, Leakey, et al. 2007). This compares to about 12 percent in modern humans. These data are significant for understanding the relative energy requirements of males and females of different hominin species (Key and Aiello 1999).

Life History of the Last Common Ancestor (LCA)

Using developmental sequences and plotting chimpanzee (*Pan troglodytes*) dental development onto a summary chart of modern human dental development, B. Smith (1991a) shows that chimpanzee dental development differs significantly from modern human dental development. Despite small sample sizes, B. Smith (1991a) is able to show that the distribution of variation of dental ages of chimpanzees differs significantly from that of modern humans, making the chimpanzee an inadequate model for human dental development (B. Smith 1991a). On the other hand, chimpanzees are the best living model for the LCA and other human ancestors (see epilogue table 1).

Life History of Basal Hominins and Early Human Ancestors

In the beginning, paleoanthropologists assumed that the developmental trajectory of extinct hominins was human-like. In his paper describing the Taung child (*Australopithecus africanus*), for example, Raymond Dart (1925, 196) stated, "The specimen is juvenile, for the first permanent molar tooth only has erupted in both jaws on both sides of the face; i.e., it corresponds anatomically with a human child of six years of age." Fifty years later research into the life history of extinct hominins continued to suggest that their development followed a human, rather than a great ape, trajectory (e.g., Mann 1975).

More than sixty years after Raymond Dart's discovery, debate over the age of the Taung child continued as Conroy and Vannier (1991) determined that the pattern of dental development was more apelike, pushing the age of Taung closer to three years old. Likewise, Alemseged, Spoor, et al.'s (2006) recent discussion of the age of another australopithecine infant, the Dikika baby (*A. afarensis*), yields a far different interpretation from that of Dart (1925). Using an African ape model to determine age at death, Alemseged, Spoor, et al. (2006) argue that their find was a juvenile. Based on the presence of fully formed, yet unerupted first molar crowns, as well as partial crowns of the second molars, and premolar and incisor crowns, Alemseged, Spoor, et al. (2006, 296) estimate the age of the Dikika baby to be about three years.

Clearly, direct counting of growth stripes to yield age at death is most desirable. When adequate samples to do this are unavailable, it is critically important to use the correct dental model (great ape or human) when estimating the age at death (and the rate of development). Because human dentition takes much longer to develop than great ape dentition, the wrong model will yield ages much older (human) or much younger (ape) than the true age.

Employing the same developmental technique described previously, B. Smith (1991a) also determined the degree of variation of dental age for several early human ancestors, including *A. afarensis* and *A. africanus*. Like chimpanzees (*P. troglodytes*), gracile *Australopithecus* fossils (*A. afarensis* and *A. africanus*) in the study also show significant variation in dental ages. Again, despite small sample sizes, B. Smith (1991a) is able to show that the distribution of variation of dental ages of gracile australopithecines differs significantly from that of modern humans, but is statistically indistinguishable from those of *P. troglodytes*. In other words, they display an apelike growth pattern (see also B. Smith 1986; B. Smith 1994).

Using this same technique, B. Smith (1986, 1994) noted that the *H. habilis* dental developmental sequence is also better approximated using great ape, rather than human, growth patterns. Contrary to her expectations, both *Paranthropus robustus* and *P. boisei* appeared to be more similar to humans

Lancaster and Lancaster's Parental Investment Scenario

In "Parental Investment: The Hominid Adaptation," Lancaster and Lancaster (1983) propose that provisioning of juvenile offspring by both parents was a major factor in the evolution of the human family and human life history.

First, they discuss the division of labor among early hominids (now called hominins). They say a dietary/ecological shift, in which the sexes specialize in acquiring different types of food, "forms the fundamental platform upon which all human behavior and elaborations of culture are built" (Lancaster and Lancaster 1983, 34). Males specialize in hunting meat, which is highly prized and unpredictable and difficult to acquire, requiring risky, dangerous behavior. Females, on the other hand, specialize in gathering plant food and acquiring insects and small vertebrates. Their endeavors are less dangerous, and the food they collect provides the majority of calories for the group. Lancaster and Lancaster argue that the success of this pattern depends upon bipedalism (and the consequent ability to carry items), a willingness to postpone consumption and share food with others, and a home base to which everyone returns.

The authors contrast this pattern with that seen in nonhuman primates, in which food sharing is limited to mothers and offspring. Consequently low-ranking, young, or sick individuals suffer during food shortages because of their inability to find, obtain, or compete for food. According to the authors, our ancestors evolved a unique solution to this problem by regularly sharing food. Food sharing allowed early hominids to eat high-quality protein on a regular basis, without suffering from the food shortfalls suffered by carnivores.

Next, the authors use data from archaeological sites to infer the history of the division of labor in hominids. They discuss possible butchery sites and home bases, suggesting postponed food consumption (i.e., butchery, but not consumption), indicating a shift from a nonhuman primate feeding pattern to a more human-like one. Since these sites date back to as far as two million years ago, Lancaster and Lancaster posit that a human-like division of labor dates back to this time.

Next, Lancaster and Lancaster discuss the division of labor in relation to feeding juveniles. The authors note that the division of labor in food acquisition benefits adults, and also provides a form of caloric "insurance" for juveniles. In nonhuman primates, once offspring are weaned, they must obtain their own food. In contrast, human juveniles are not expected to find enough food on their own to sustain themselves for several years after weaning. The authors note that this, rather than the foods that are eaten, is an important contrast between humans and nonhuman primates because it requires "collective responsibility" for obtaining and sharing food (1983, 36).

The significance of this difference in "collective responsibility" is obvious when comparing the survivorship curves of human hunter-gatherers with other mammals. The proportion of juveniles who survive to adulthood is much higher in hunter-gatherers (46 to 58 percent) than in free-ranging nonhuman primates (13 to 38 percent). The authors posit that this difference in offspring survival (due to provisioning) may explain why the extremely successful apes of the Miocene declined sharply in the Pliocene and Pleistocene, while hominid populations increased.

Humans have long life spans and prereproductive periods, even compared to other primates. Studies of life history strategies across species indicate that increases in both these parameters require access to predictable resources. Lancaster and Lancaster suggest that these predictable resources may have come from enhanced tool-using abilities for extractive foraging while continuing offspring provisioning.

Once Lancaster and Lancaster establish that a division of labor in food acquisition enables the long-term provisioning of juveniles, they address the evolution of the role of "father" in hominid life history. They say the first step in food sharing probably occurred between mother and offspring, but extensive provisioning of offspring seen in humans requires a division of labor and combined effort by both sexes. Once males began sharing responsibility for feeding their offspring, a new set of attachments likely developed. The role of males in human society, as "father-husband," with all of its economic and social obligations, is unparalleled in nonhuman primates (1983, 43).

Lancaster and Lancaster next examine the life cycle of human females from the perspective of parenting behavior, noting that its major features support high levels of long-term investment in multiple young of different ages. Based on data from females in hunting-gathering populations, which more closely approximate the lifeways of our hominid ancestors, these features include a long period of adolescent sterility, a late age at first birth, nearly continuous nursing during the reproductive phase of adulthood, long interbirth intervals (about four years), low fertility rates, low frequency of menstruation, and menopause before death (1983, 43).

Late age at first birth and adolescent sterility preceding the first birth ensure that females are fully physically and mentally mature before they have their first child. Thus, their investment in their first child does not compete with investment in their own biological, social, or mental development, promoting greater success in child rearing. The authors point out that among hunter-gatherers, mothers invest heavily in their offspring during the first four years. Such high levels of care are possible because of the long interbirth intervals and small numbers of children born among hunter-gatherer societies. Long birth intervals appear to be related to diet, frequent, prolonged lactation, and activity patterns on the part of the mother, although infanticide may also play a role (Lancaster and Lancaster 1983, 48).

Finally, the authors turn to female menopause, discussing it from the perspective of parental investment. As such, female menopause is adaptive when the long-term dependence of juveniles requires extensive care on the part of the mother. Ending reproduction early (i.e., via menopause) helps ensure that the last child will be cared and provided for until it can survive on its own. In conclusion, Lancaster and Lancaster argue that intense parental investment was so important in hominid evolution that it shaped basic features of human reproductive biology and the human life cycle.

than gracile australopithecines in their dental development (B. Smith 1986, 1994).

Counting incremental growth lines visible on the surface of the incisor crowns, Bromage and Dean (1985) determined that the incisor crowns of three young australopithecines (SK 63, STS 24, and LH 2), which died near the time of the eruption of M_1, formed around three years of age. These ages match the age for the eruption of M_1 in chimpanzees (3.3 years; Nissen and Riesen 1964), and differ greatly from the typical age of M_1 eruption in humans (5.5 to 7 years; Dahlberg and Menegaz-Bock 1958; see also B. Smith 1991a; table 6.3).

Based on their results, Bromage and Dean (1985) also proposed that early *Homo* matured as quickly as *Australopithecus*. This has yet to be confirmed, since there are no early *Homo* fossils from individuals who died near the time of M_1 eruption. Because incisor growth bands are best used to calibrate growth that occurs during the development of incisor crowns (M. Dean 1987; B. Smith 1994), statements regarding the maturation of early *Homo* require fossil material from individuals who died at a relatively early age.

Life History of Middle-Period Human Ancestors

Life history in *H. erectus* (and *H. ergaster*) and archaic humans has also been diagnosed by comparing developmental dental sequences. Plotting *H. erectus* dental development onto a summary chart of modern human dental development, B. Smith (1991a) shows that the *H. erectus* specimen approximates the human profile to a much higher degree than the chimpanzee profile. Comparing the dental ages reflected in the different teeth, B. Smith (1991a) shows that the ages of *H. erectus* vary only slightly more than the modern human sample, resulting in a much lower degree of variation of dental ages. She says this means that the sequence of *H. erectus* dental development is more human-like than chimplike (see also B. Smith 1994). The degree of variation of dental age for the dentition of the lower jaw of *H. erectus* is within the range of that found in modern humans (B. Smith 1991a).

More recent work on the dentition of Turkana Boy specimen number KNM-WT 15000 (*H. ergaster*) and *H. erectus* from Sangiran, Java, confirms that the *sequence* of dental events in these species was like those of modern humans. However, contrary to earlier models, C. Dean, Leakey, et al. (2001) say that the timescale of the dental development of these fossils differed from that of humans because the underlying enamel processes differed. Based on their estimate of crown formation times, they reconstruct that M_1 emerged at 4.4 years and M_2 at 7.6 years in *H. erectus*. Consequently, they estimate that the Turkana Boy was closer to eight than twelve years of age at death. "If correct, these estimates of molar emergence times have shifted a little, in step with brain size from those known for African great apes and australopiths. Nevertheless, it now seems increasingly likely that a period of development truly like that of modern humans arose after the appearance of *H. erectus*, when both brain size and body size were well within the ranges known for modern humans" (C. Dean, Leakey, et al. 2001).

Life History of Late-Period Human Relatives

Archaic *H. sapiens* and Neanderthals show a more human-like dental development pattern. By comparing the sequence of dental formation in three Neanderthal specimens (Gibraltar 2, Teshik Tash, and Ehringsdorf), to both human and ape dental development profiles, B. Smith (1994) determined that archaic *H. sapiens* and Neanderthals have a human-like, rather than apelike, dental growth pattern (see also B. Smith 1991a; B. Smith and Tompkins 1995). C. Dean, Leakey, et al. (2001) confirm this.

Correlations between Dental Development, Longevity, and Brain Size

Comparative studies of mammalian life history have shown that slow maturation and long life are associated with increased brain and body size, enhanced learning and sociality, and greater parental care (e.g., Sacher 1959; Pianka 1970; Sacher 1982; Harvey and Clutton-Brock 1985; Lillegraven, Thompson, et al. 1967; Charnov and Berrigan 1993; Leigh and Blomquist 2007). Humans clearly fit this pattern, having both a large relative brain size and long maximum life span compared with other mammals (Sacher 1975).

As discussed earlier, humans fall at the extreme end of the continuum in most life history variables, with longer maximum life spans, later ages at puberty, and later ages at first reproduction than our closest living relatives, the great apes (table 6.2; see also Sacher and Staffeldt 1974; B. Smith and Tompkins 1995). Whereas large brains allow enhanced behavioral flexibility and complexity, they entail high energy costs (R. Martin 1981; E. Armstrong 1983). The large (relative to body size) human brain, for example, uses 20 percent of the adult basal metabolic rate, compared to 4 to 6 percent for the smaller-brained rat, cat, and dog (E. Armstrong 1983; Hoffman 1983). The neonatal human brain uses a whopping 80 percent (Lancaster 1986).

In order to bear these costs, the associated benefits must have been even greater. These benefits include extended parental care and an increased period of immaturity, which provide time to learn (see also B. Smith and Tompkins 1995). Several studies of brain size, metabolism, and life span indicate a close link among the three: that brain metabolism determines the rate of growth and aging in vertebrates and that relative brain size controls maximum life span (Sacher 1959, 1975; Sacher and Staffeldt 1974; see also Hoffman 1983 and E. Armstrong 1983). Thus, we expect to find large brains and long life spans in K-configured mammals, particularly among primates, who also invest heavily in relatively few offspring (table 6.1; R. Martin 1983; see also B. Smith 1989).

Key and Aiello (1999) argue that larger brains and the shift to an animal-based diet in *H. erectus* led to a shift in the relative energy requirements of male and female. As sexual dimorphism decreased, female energy requirements increased both absolutely and relatively to those of males. They believe these changes would have provided strong selection pressures for provisioning and other parental investment by males (as indicated above, these creatures may have been more dimorphic than previously thought).

In primates, age at eruption of M_1 and M_3 is highly correlated with body weight, birth weight, gestation length, age at weaning, age at puberty, age at first breeding, maximum life span, neonate brain weight, and adult brain weight (B. Smith 1989). Therefore, late age at eruption of M_1 and M_3 indicates larger values and later ages for these other life history variables. Hence the age at eruption of M_1 and M_3 provides indirect evidence of length of immaturity and the degree of parental investment or individual learning, provided investigators use the correct model (ape or human).

Given the strong link among large brains, increased life span, and periods of immaturity, it is likely that the human pattern of life history evolved in conjunction with brain size (B. Smith and Tompkins 1995). These links, however, do not tell us what brain sizes would be associated with what developmental profiles in extinct hominin species. Dental studies of various earlier hominins are beginning to give us some ideas about this. Surprisingly, the redating of *H. ergaster/erectus* suggests that 900 to 1,000 cc is inadequate, and only the 1,250 to 1,400 cc of archaic and modern *H. sapiens* seems adequate to entail a modern life history.

To summarize, recent research indicates that dental development in *A. afarensis*, *A. africanus*, *H. habilis* (Bromage and Dean 1985; B. Smith 1986; Conroy and Vannier 1991; B. Smith 1991a), and even in *H. ergaster/erectus* (C. Dean, Leakey, et al. 2001) is more similar to the relatively rapid ape trajectory than to the slower, human trajectory. It is not until the appearance of *H. neanderthalensis* that we see clear signs of the slower, more human pattern of dental development (B. Smith 1991a, 1994; B. Smith and Tompkins 1995).

This research implies that early members of the hominin lineage developed rapidly compared to modern humans. Consequently, the dental data imply that they would have had lower values of associated life history variables: adult body size, neonate body size, gestation length, age at weaning, age at puberty, age at first breeding, maximum life span, neonate brain weight, and adult brain weight (for brain size and body size, we have independent data from the fossil record). Accordingly, *A. afarensis*, *A. africanus*, and even *H. erectus* would have had shorter periods of immaturity with reduced time for learning and therefore less behavioral flexibility and complexity than modern humans. Given the new data, it now seems highly likely that the pattern of modern human life history evolved after *H. erectus* and only with archaic *H. sapiens* (C. Dean, Leakey, et al. 2001).

Longevity and Reproductive Strategies

Based on the brain size–life span link proposed by Sacher (1975), O'Connell, Hawkes, et al. (1999) suggest that *H. erectus*, with a cranial capacity of 800 to 1,100 cc (intermediate between chimpanzees and modern humans), had a maximum life expectancy intermediate between chimpanzees (fifty years) and modern humans (ninety to a hundred years). Surprisingly, recent dental analysis does not support this interpretation (C. Dean, Leakey, et al. 2001).

Dental development in *H. neanderthalensis* indicates a slower rate of development than australopithecines, *H. habilis* (B. Smith 1991a, 1994; B. Smith and Tompkins 1995), and even *H. erectus* (B. Dean, Leakey, et al. 2001). This suggests a lengthened period of immaturity during which greater learning could take place. It also suggests an increased maximum life span. Whereas *H. neanderthalensis* appears to display a slower, more human-like developmental trajectory, however, their maximum life span was short compared to that of modern humans.

O'Connell, Hawkes, et al.'s Grandmothering and the Evolution of *Homo erectus*

O'Connell and colleagues (O'Connell, Hawkes, et al. 1999) propose the "grandmother hypothesis" to explain the evolution of human life history in *Homo erectus*. They suggest that climate- and environment-driven changes in female foraging and food sharing, especially the exploitation of tubers, were responsible for the evolution, distribution, and persistence of *H. erectus*.

They begin by reviewing typical inferences about this species, that is, the appearance of nuclear families, central place foraging, and a sexual division of labor whereby males hunt and provision females and offspring. They note that this pattern in *H. erectus* has been "explained by a long-term trend toward cooler, drier climates that reduced the availability of previously important plant foods while favoring the spread of game-rich savannas" (O'Connell, Hawkes, et al. 1999, 463). Increased reliance on big-game hunting by males meant that females paired with males to ensure their access to resources, and, as a result, could reduce their own foraging efforts. In turn, female fertility and offspring survival improved; this pattern resulted in an extended period of juvenile dependence, increased cranial capacity and learning, and greater behavioral flexibility.

However, the authors point out, this traditional scenario for *H. erectus* has been challenged in recent decades on three fronts. First, research on chimpanzees indicates that while males engage in hunting, there is no evidence for central place foraging or male provisioning of females. Second, research on modern hunter-gatherers indicates that big-game hunting serves purposes other than provisioning wives and children (i.e., showing off). Third, archaeological evidence indicates that *H. erectus* engaged more regularly in scavenging than killing large game, so that males rarely acquired large amounts of meat for provisioning females and offspring.

Maternal provisioning of dependent offspring is central to O'Connell and colleagues' alternative "grandmother hypothesis" for explaining for the

---

Research on the age at death of adult Neanderthals indicates that their maximum life span was fifty years (T. Stewart 1977; D. Thompson and Trinkaus 1981; Trinkaus and Thompson 1987; Trinkaus 1995), close to the average age at which modern human females enter menopause (Lancaster and King 1985). It is unlikely, however, that Neanderthals reached their reproductive life span (Trinkaus and Thompson 1987), owing to high levels of stress and high frequency of injuries they suffered (Trinkaus 1983; Berger and Trinkaus 1993).

ecology and evolution of *H. erectus*. They argue that among modern gatherer-hunters, such as the Hadza, maternal provisioning allows exploitation of "habitats from which they would otherwise be excluded if, as among other primates, weanlings were responsible for their own subsistence" (1999, 466). By provisioning weaned, but dependent offspring, other individuals, often grandmothers, influence mothers' birth spacing, allowing them to reproduce again sooner. Reduction in the mothers' birth interval in turn increases the grandmothers' fitness, especially in postmenopausal women whose fertility has declined. Although human females have long life spans and juvenile periods, their annual fecundity is higher than anticipated. The authors attribute this peculiar combination of life history variables to provisioning by grandmothers. The grandmother hypothesis applies if females remain in their natal groups so grandmothers help their daughters. If grandmothers provision their *daughter-in-law's* offspring, they cannot be sure that the children are their sons' offspring.

Next, O'Connell and colleagues apply the grandmother hypothesis to *H. erectus*. First, they say life history data support the idea that *H. erectus* displayed longer life spans and later maturity than earlier hominids (now called hominins). The authors suggest that life history changes were prompted by a decline in the availability of easily obtained resources (e.g., fruit) when climates became cooler and drier. Under such conditions, the authors suggest *H. erectus* would have had two choices: to change their foraging patterns to continue exploiting "'children's resources" (1999, 470), or begin exploiting new, widely available resources difficult for young individuals to access (i.e., underground storage organs, or tubers). Because exploiting underground storage organs requires increased foraging by adults, grandmothers could have increased their daughter's (and their own) fertility, by provisioning their grandchildren. In turn, the ability of *H. erectus* to exploit this food resource efficiently would explain widespread distribution of *H. erectus* in dry habitats previously unexploited by hominids.

Given these new studies, it seems likely that the modern human life history pattern evolved very late, probably in the past two hundred and fifty thousand to one hundred and sixty thousand years. In the following sections, we discuss three hypotheses proposed to explain the evolution of this pattern: Lovejoy's (1981) pair-bonding hypothesis, O'Connell, Hawkes, et al.'s (1999) grandmother hypothesis, and Lancaster and Lancaster's (1983) parental investment hypothesis.

LIFE HISTORY SCENARIOS

In his widely cited pair-bonding scenario, Owen Lovejoy (1981) attempts to explain the evolution of human characteristics, including bipedalism and large brain size. Noting the correlation among long maximum life span, longer gestation, longer weaning, fewer offspring, and increased interbirth intervals, he says that females with a longer life span and delayed reproduction would have had to survive longer to maintain the same reproductive output as individuals who reproduced earlier and more frequently, or their mortality rate would have had to have been lower (Lovejoy 1981).

According to Lovejoy (1981), traits associated with increased life span, including increased intelligence, higher levels of parental investment, and longer periods of learning would have reduced environmentally induced mortality (predation, accident, disease, etc.). Furthermore, he proposes the evolution of monogamous pair bonds, in which males foraged farther afield than females (reducing female energy expenditure) and provisioned females and offspring.

He argues that provisioning females and offspring would increase males' reproductive success (as long as paternity certainty is high) by increasing the survivorship of his offspring and reducing the interbirth interval of his mate (Lovejoy 1981). Provisioning (and other potential forms of paternal care) would increase offspring survival, as young, dependent individuals required more care and more time to develop and learn.

According to the hypothesis proposed by Kristen Hawkes, James O'Connell, and colleagues (Hawkes, O'Connell, et al. 1998; O'Connell, Hawkes, et al. 1999; Hawkes 2003; Hawkes 2004), human females evolved menopause and long postreproductive life spans because "grandmothers" shared food with their daughters' children. Older females, without their own dependent offspring, could increase their reproductive success by helping their daughters feed their grandchildren. Aid to reproductive-age females allowed them to shorten their interbirth interval (Hawkes, O'Connell, et al. 1998; Hawkes 2004). This favored the evolution of a long postreproductive period in human females (Hawkes, O'Connell, et al. 1998).

O'Connell, Hawkes, et al. (1999) suggest that the human life history pattern, particularly the long postreproductive period for females, evolved with *H. erectus*. They think this occurred in relation to declining resources easily obtained by young children (e.g., fruit) when climates became cooler and more seasonal. According to the authors, adult *H. erectus* provisioned their children with alternative resources: subterranean tubers (underground storage organs, or

USOs). Such food sources were widely available in cooler, seasonally dry habitats in which *H. erectus* found themselves (see O'Connell, Hawkes, et al. 1999). USOs would have been difficult for youngsters to utilize because they require preconsumption processing in the form of baking, boiling, leaching, or roasting (see chapter 4, on subsistence, for another USO scenario).

The ability to process tubers allowed *H. erectus* to take advantage of an abundant resource previously little utilized by hominins. Thus, according to the authors, female foraging may have become an important part of *H. erectus*'s resource acquisition. Because processing of this food source is difficult or impossible for young children, adults (i.e., mothers) would have had to process and provision their offspring with these foods. Once food sharing between mothers and children became widespread, older, nonreproductive females could increase their reproductive success by helping their grown daughters provide for their offspring (Hawkes, O'Connell, et al. 1998).

Lancaster and Lancaster (1983) propose another food-sharing scenario based on the model of a division of labor between hunting males and gathering females, with delayed consumption and transport for provisioning young. They note that long-term postweaning food sharing with juveniles is unique to humans among primates, resulting in much higher survival to adulthood (46 to 58 percent) in hunter-gatherers than in free-ranging nonhuman primates (13 to 38 percent). They suggest that the first step in food sharing was between mother and offspring, followed by paternal sharing, leading to attachment. They argue that a long period of adolescent sterility (or subfertility [Hrdy 1999]) allowed females complete growth and sufficient fat reserves for prolonged lactation to support early infant brain development (Lancaster 1986). This led to later onset of reproduction and longer birth intervals. The authors suggest on the basis of early butchery sites that the division of labor began as early as two million years ago.

CONCLUSIONS ON HUMAN LIFE HISTORY

The various scenarios discussed here suggest that increased parental or kin investment in females and offspring were the keys to the emergence of a distinct human childhood, adolescent sterility (under hunter-gatherer conditions), and menopause, along with increased brain size. In each case food sharing is related to a particular subsistence strategy: gathering, hunting, or excavating USOs. We believe that a mixed strategy of gathering and excavating foods, supplemented by scavenging and hunting, is the most likely scenario (see chapter 4).

The scenarios agree that help from other individuals significantly increased the reproductive success of hominin females. In Lovejoy's (1981) pair-bonding scenario, help comes from her mate, who provisions her and her offspring, enabling longer periods of immaturity and time to learn, at the same time reducing the interbirth interval and decreasing mortality. In O'Connell and Hawkes's (Hawkes, O'Connell, et al. 1998; O'Connell, Hawkes, et al. 1999; Hawkes 2003, 2004) grandmother scenario, help comes from the mother's mother, increasing her reproductive success by allowing shorter birth intervals. In Lancaster and Lancaster's (1983) scenario, increased parental investment in juveniles comes from both parents, allowing increased reproductive success through increased offspring survival (we should note that pair bonding and shared parental care need not imply monogamy [Daly and Wilson 1983]; see chapter 7). Any or all of these factors could have driven the evolution of human life histories.

The proposed timing of these scenarios, however, is questionable. Lovejoy's (1981) scenario is vague about which species first displays changed life history patterns, although it focuses on *A. afarensis*. Lancaster and Lancaster (1983) propose an early beginning for food sharing, about two million years ago. O'Connell, Hawkes, et al.'s (1999) scenario, in contrast, suggests that changes in life history occurred about 1.6 million years ago with the appearance of *H. erectus*, with its increased cranial capacity and ability to use fire. Likewise, Key and Aiello (1999) argue that reduced sexual dimorphism and increased brain size associated with a meat diet would have favored male provisioning in *H. erectus*.

The timing in all these scenarios is *inconsistent* with new dental development data, which suggest that longer, more human-like periods of development and maximum life span began after *H. erectus*. These data imply that a final shift in life history occurred much later with the emergence of archaic *H. sapiens*. Nevertheless, changes in *H. erectus* must have provided the platform for further selection on life history.

Taken together, the life history data reviewed in this chapter suggest two major stages in the evolution of human life history: first, a great ape–like pattern in basal and early hominins continuing into early *Homo* and even *H. erectus*, finally changing into the modern human pattern coincident with the emergence of modern *H. sapiens*, possibly as late as two hundred and fifty thousand to one hundred and sixty thousand years ago. This picture is consistent with the late evolution of the modern human brain, and the late evolution of language (see chapter 8), higher intelligence (see chapter 9), and cultures (see chapter 10).

7

Origins of Human Bodily Displays

The view which seems to me the most probable is that man, or rather primarily woman, became divested of hair for ornamental purposes, as we shall see under Sexual Selection, and according to this belief, it is not surprising that man should differ so greatly in hairiness from all other Primates.

—*Darwin 1871, 439*

This chapter deals with the functional significance of bodily displays, that is, movements of various body parts specialized for sending signals to other members of the same species. Intention movements, protective movements, feeding, thermoregulating, respiratory, and elimination acts provide raw materials for signals. These movements have evolved into displays through a process of *ritualization* (J. Huxley 1942). Structures involved in signaling movements have been specialized through enlargement, contouring, patterning, and bright coloration. Their movements have been specialized for freezing, repetition, typical sequence, and orientation toward receivers (see summary in Eibl-Eibesfeldt 1979).

Signals have evolved because they have enhanced the survival or reproductive success of signalers, but not necessarily receivers. They are more akin to advertising than to unbiased information transfer (Dawkins and Krebs 1978). Signals are more energy efficient than other behaviors, for example, ritualized threat displays may intimidate opponents without escalating to physical confrontation. In courtship, individuals use ritualized displays to attract mates. Receivers use signals to assess the signaler's potential quality as a parent. In some bird species, for example, courting males bring food to the female (Dawkins and Krebs 1978). This ritualized display indicates the male's ability to feed the chicks.

To avoid being deceived, recipients must look for "honest" signals, that is, those costly to perform. When displays are energetically expensive to produce, they reveal something useful (e.g., health, reproductive potential). As they grow larger and more complex, they become more difficult to fake. Such traits have been called *fitness indicators* (Miller 2000). In contrast, some displays grow larger and more exaggerated through *runaway selection*. Although members of the opposite sex prefer such traits, they do not necessarily reveal anything about the signalers' fitness. In runaway selection, traits have become more pronounced because individuals of the opposite sex prefer evermore exaggerated forms of the trait (Eberhard 1985). These supernormal stimuli take advantage of preexisting nervous system biases (West-Eberhard 1983).

After a brief discussion of sexual selection theory, this chapter focuses on the evolutionary significance of various uniquely human bodily displays. We begin with discussion of skin, body hair, and facial features, common to both sexes. Then we treat other bodily displays as male or female displays designed to attract members of the opposite sex or intimidate members of the same sex. Because there are few clues regarding the evolution of the bodily displays, we cannot trace their evolution through stages of hominin evolution. Rather, we focus on the evolutionary significance and function of bodily displays in modern humans. Finally, we attempt to trace the appearance of a few bodily displays during the course of hominin evolution.

SEXUAL SELECTION THEORY

Throughout this book, we have used Darwin's (1859) theory of natural selection to explain the evolution and functional significance of uniquely human characteristics. In this chapter, we focus on Darwin's (1871) theory of sexual selection, which explains traits that do not increase individuals' ability to survive (or, worse, hinder survival). A classic example of this is the heavy and cumbersome peacock's tail, which Darwin (1859) suggested was an "inconvenience." Indeed, the large, brightly colored tail of the peacock makes males more visible to predators and hampers escape. He noted that peahens prefer to mate with males displaying a vibrant tail. Large, vibrant tails are *fitness indicators* because only healthy peacocks can muster the energy required to grow them. By choosing males with the largest, brightest tails, females choose the healthiest mates. The importance of mating cannot be overstated. If a trait improves chances of successful mating, *and* it has a genetic basis, it will increase in frequency in the population, even if it is costly in terms of survival.

The most basic difference between the sexes is that males produce small gametes (sperm), while females produce large gametes (eggs). Whereas male and female gametes contain DNA, only eggs contain energy for the embryo. Females invest more energy in a single gamete than males do. Given limited energy, the greater energy they invest in eggs means that compared to males, females produce relatively few gametes (Trivers 1972).

In addition, in species with internal fertilization, females invest more in gestation and parental care than males. Female mammals spend a significant part of their lives pregnant or lactating and therefore unable to conceive. Thus, in any given population fewer females than males are available for mating (Houde 2001). Consequently, males and females have evolved different reproductive strategies.

In general, female mammals increase their reproductive success by producing viable offspring and investing in each one. Males, on the other hand, maximize their reproductive success by mating with as many females as possible (Bateman 1948). Although male parental investment varies widely from species to species, depending upon confidence in paternity and mating system, it is always less than female investment (Trivers 1972). Because more males than females are available for mating, females are a limiting resource for males (Bateman 1948). Therefore males generally benefit from competing with other males for access to females. They also benefit from controlling females whenever possible. Female choice is inversely related to male control (Borgia 1979; see table 7.1). On the other hand, because they have limited gametes, females generally benefit by carefully choosing mates. Hence, male competition and female choice are the foundations of sexual selection theory (Darwin 1868).

Members of the sex that compete among themselves (among mammals, often, but not always, males) develop characteristics that enhance their competitive

Table 7.1. **Degrees of Female Choice**

	High ←——— Degree of Female Control in Mate Choice ———→ Low			
	Unrestricted Female Choice	Males Collect Resources	Males Control Resources	Males Control Females
Definition	Females have free access to resources and choose among males on the basis of genetics.	Females have access to resources, but males exchange resources they have collected for matings. Females are free to choose among males, but may sacrifice genetic quality to gain access to better resources if the genetically superior male does not have the best resources.	Female access to resources is through males. Females choose males based on the resources they hold. Female choice is severely limited, especially if a few (or only one) males control all of the resources.	Females have no choice because males capture females and do not allow them to mate with other males. Capture denies females the ability to choose males for the resources they provide or based on genetic quality.
Examples	Lekking species of birds.	Chimpanzee males trading hunted meat for sex with females.	Bird species in which males set up territories around nest sites females need to breed.	Male hamadryas baboons "kidnapping" juvenile females.

Source: Adapted from Borgia, G. 1979. Sexual selection and the evolution of mating systems. In *Sexual selection and reproductive competition in insects,* ed. M. S. Blum and N. A. Blum, 19–80. New York: Academic Press.

abilities. These can include large body size, large canines, antlers, and heightened aggression. Members of the sex that compete for the interest of the other sex (among birds and mammals, often, but not always, males), evolve characteristics that increase their attractiveness. These can include brightly colored, enlarged organs, and associated behaviors (Darwin 1868; Crook 1972; Selander 1972).

Nevertheless, Darwin (1871) recognized that both males and females may compete and choose, in modern terms: (1) *intra*sexual selection favors traits that increase the ability of members of one sex to compete among themselves for access to the other sex, and (2) *inter*sexual selection favors traits that increase the ability of one sex to attract the other. Both patterns can occur in both males and females, resulting in four distinct categories of sexually selected traits (table 7.2). As Darwin observed, human sexes are subject to "double selection."

The degree and kinds of intrasexual competition and intersexual choice correlate with mating systems. The sex ratio (number of males to number of females) in most species is close to 1:1, meaning that there are nearly equal numbers of males and females. In *polygynous* (harem) mating systems, some males mate with many females, and many males do not mate. In this mating system, male-male competition for access to groups of females is intense, and males are much larger than females. In some polygynous species, like gorillas and orangutans, males are twice the size of females.

In *polyandrous* mating systems, one female mates with many males, the situation is reversed, and some females have no mates. In this mating system, female-female competition for access to mates is intense, and females tend to be larger than males. Tamarins and marmosets are primate species with a polyandrous mating system.

Table 7.2. **Pathways to Sexually Selected Traits**

Intrasexual Selection	Examples	Intersexual Selection	Examples
Male-male competition	Physical fighting between males, sperm competition	Female choice	Size and shape of the male penis
Female-female competition	Reproductive suppression (e.g., tamarins, marmosets)	Male choice	Large breasts, large hips

In *polygamous* mating systems (sometimes called *promiscuous* mating systems), many males mate with many females, and males compete for copulations. In polygamous species, like chimpanzees, sexual dimorphism is less extreme, and males are only slightly larger than females. Finally, in *monogamous* mating systems, one male mates with one female, so that most members of each sex have access to a mate. In this mating system, same-sex competition is greatly reduced and the sexes are similar in body size, or *monomorphic*. Gibbons and siamangs are monogamous primate species.

Because sexually selected traits usually appear at puberty in one sex or the other, adult males and females of the same species may be *sexually dimorphic* (differ in size and appearance). Males and females of polygynous species are highly sexually dimorphic. Their secondary sexual characteristics (e.g., enlarged breasts and hips in human females, shoulder and arm muscles in human males) have been shaped through sexual selection (Darwin 1871). These characteristics appear at puberty under the influence of sex hormones. Genitals, primary sex characteristics, also enlarge at puberty.

If these sexual characteristics are tied to overall quality and are energetically expensive to produce, they may become more exaggerated because they are good fitness indicators and signal important information about the health and/or reproductive status of the individual (Miller 2003). As we will see, traits such as facial symmetry in males and the hourglass figure in females are good examples of human fitness indicators.

HUMAN BODY DISPLAYS

Loss of Body Hair

Much has been made of our "hairlessness" (see Darwin 1871), but humans have more hairs than chimpanzees. Each hair is relatively short, so we appear to have little or no body hair. Therefore, as Desmond Morris (1967) points out in *The Naked Ape*, functionally, humans are "naked" as our skin is visible and unprotected by fur. We discuss four hypotheses proposed to explain "hairlessness" in humans: Darwin's (1871) sexual selection hypothesis, the thermoregulation hypothesis (Wheeler 1984, 1991, 1992), the aquatic ape hypothesis (Hardy 1960; Morgan 1982), and the parasite load hypothesis (Pagel and Bodmer 2005).

Alister Hardy's Aquatic Scenario

In his article, Alister Hardy (1960) proposes that in the Miocene, a "branch of a primitive ape-stock" went through an aquatic phase during which they acquired certain distinctively human characters: He suggests that they were "forced by competition from life in the trees to life on the sea-shores to hunt for fish, shell fish, sea-urchins, etc. in the shallow waters off the coast" (642).

Hardy argues that this aquatic history could explain a number of features including Man's "exceptional ability" to swim, his loss of body hair, and subcutaneous fat, which are typical of such aquatic mammals as whales, manatees, and hippopotamuses (1960, 643). He argues that this history also explains Man's erect posture and tool-making hand. Regarding hair loss, he says, "Man has lost his hair all except on the head, that part of him sticking out of the water as he swims; such hair is possibly retained as a guard against the rays of the tropical sun, and its loss from the face of the female is, of course, the result of sexual selection" (643). He also suggests that the direction of growth of body hair, which meets on the midline, arose as a means for streamlining hair flow during swimming before hair loss occurred. Regarding the evolution of erect posture, Hardy says that toddling on the shore hunting and then swimming led to intervals of resting, feet on the bottom, head out of the water, "in fact, standing erect with the water supporting his weight he would have to raise his head out of the water to feed; with his hands full of spoil he could do so better standing than floating" (645). He notes that he imagines Man sleeping on shore. Finally, he proposes that Man learned to use tools on the shore "to crack open the enshelled 'packages' of food which were otherwise tantalizingly out of his reach" (645). Regarding the timing, he says, "It is in the gap of some ten million years, or more, between *Proconsul* and *Australopithecus* that I suppose Man to have been cradled by the sea" (645).

Darwin (1871) argued that the "lack" of human body hair was a sexually selected characteristic because men prefer hairless women, and because "in all parts of the world, women are less hairy than men." Darwin concludes that because females are less hairy than males, females were the first to lose their body hair. In doing so, he argues, these females passed on their trait for hairlessness equally to their male and female offspring.

Wheeler (1984) argues that human hairlessness evolved to enhance the body's ability to cool itself. He says hominins lost body hair when they began occupying open savanna environments, in which they experienced increased sun exposure. Wheeler's hypothesis is that bipedality evolved, at least in part, to reduce sun exposure by reducing the surface area that would be in direct sunlight during midday (Wheeler 1984, 1991, 1992). In a 1992 paper, he suggested that in combination with upright posture, reduced body hair enabled hominins in savanna environments to radiate heat back into the environment or dissipate heat more easily.

The aquatic ape hypothesis, first proposed by Alister Hardy (1960), and elaborated by Elaine Morgan (1982), is that our ancestors experienced an aquatic or semiaquatic lifestyle. According to Elaine Morgan, loss of body hair and increased levels of body fat evolved in these "aquatic apes" because fur is a poor insulator under water.

Pagel and Bodmer's (2005) parasite load hypothesis (which incorporates both natural selection and sexual selection) expands on the old idea that loss of body hair is related to reduced parasite loads (Darwin 1871). They say that because hairlessness reduces parasite loads, reduced body hair would have increased individual survival and reproduction in our ancestors. They argue that this adaptation could only evolve after our ancestors had developed the ability to protect their "naked" skin from exposure to heat, cold, and rain.

Once hairlessness was established, Pagel and Bodmer (2005) argue, sexual selection enhanced the trait because it advertised reduced parasite loads. Unlike other sexually selected traits, hairlessness, under this hypothesis, would have been a desirable trait in both sexes since it signaled enhanced health. The authors suggest that the greater loss of body hair in females is due to the "stronger sexual selection from male versus female mate choice in humans" (S118).

As Dale Guthrie (1976) points out in *Body Hot Spots*, humans have thicker hair than the apes in the underarms and genital regions, as well as on top of the head and on male faces. According to Guthrie, these strategically placed tufts of hair are devices for dispersing scent. In many mammals, scent is an important sexual and/or threat signal. Because humans in Western society cover their natural scent with soap, deodorant, and perfume, we may think scent is unimportant. However, humans have more, and larger, scent-producing glands (sebaceous and apoecrine) than other hominoids (Stoddart 1990;

Wheeler's Temperature Regulation Scenario

P. E. Wheeler (1984) argues that bipedalism arose as an adaptation for temperature regulation rather than tool use, as some have proposed. He begins by noting that large-brained primates experience overheating in hot environments because they lack certain physiological adaptations that allow other mammals to selectively cool their brains. These include evaporation from the mucus lining of the nasal turbinates and a special net of blood vessels in the cavernous sinus that cools blood going to the brain. For this reason, he says, "one of the principal adaptations required by these primates to exploit fully the more open savannah environments would have been as extremely effective whole-body cooling system" (92).

He argues that hominids (now called hominins) solved the thermoregulatory problem by adopting upright posture. Bipedality reduces "the fraction of the total body surface actually presented to the incoming rays when the sun is directly overhead by 40% as compared to that experienced by quadrupeds" (Wheeler 1984, 93). It therefore also reduces the need for body hair to shield the body. He postulates that bipedalism preceded the loss of body hair.

Regarding hair loss, he says, "Bipedality also alleviates the other major problem associated with naked skin. It greatly reduces the areas presented perpendicularly to the most intense fluxes and therefore the highest risk of thermal damage. In a bipedal hominid these areas are confined to the head and upper shoulders and can be protected by retaining a small area of hair cover" (1984, 94). Naked skin reduces loss of body hair and the layer of air it traps results in increased thermal conductance. This combined with increased evaporation from sweating facilitates heat loss. Subcutaneous fat can compensate for heat loss at night when temperatures are low.

According to Wheeler's evolutionary scenario, the first-stage ape ancestor was semierect with dense body hair and a 400 cc brain, his hominid descendant *Australopithecus afarensis* was bipedal with dense body hair and a 400 cc brain, and his descendant was bipedal with naked skin and a 400 cc brain. The most recent descendant was bipedal with naked skin and a 1,400 cc brain. He argues that the thermal stability achieved by bipedalism and naked skin may have allowed subsequent brain expansion by removing physiological constraints of thermoregulation.

Pawlowski 1999). In addition, sebaceous and apoecrine glands begin functioning at puberty, which indicates that their function is associated with reproduction (Stoddart 1990).

Many human scent-producing glands are found in the underarm and genital regions. According to Guthrie, the tufts of hair found in both regions trap and disseminate scent (both regions are easily disturbed by bodily movement). They may send important sexual signals to members of the opposite sex. Singh and Bronstad (2001) have shown that males prefer the body odor of ovulating females to that of nonovulating females, while Doty, Ford, et al. (1975) have shown that males perceived odor of vaginal secretions as "pleasant" only during the ovulatory phase of the menstrual cycle. This research indicates that males may (unconsciously) use female scent as a means for assessing reproductive state.

Facial Features

Primate facial muscles and organs, originally evolved for primary functions of eating, breathing, smelling, looking, hearing, and vocalizing, became secondarily specialized for signaling (Andrew 1963). Consequently, most monkeys and apes have expressive faces. Humans and our closest living relatives, African apes, display some similar facial expressions of emotions (Goodall 1968, 1986). See figure 7.1 for a depiction of chimpanzee facial expressions.

Humans also display several uniquely human expressions, including crying with tears, smiling, and blushing. In *The Expression of the Emotions in Man and Animals*, Darwin (1965, 1998) concludes from his cross-cultural survey that human facial expressions are universal. He argues that some expressions evolved from movements designed to increase information flow or to protect the eyes, nose, or mouth from noxious stimuli. He notes, for example, that because disgust arises in connection with eating, it is mainly expressed by protective movements of the mouth: "Moderate disgust is exhibited in various ways; by the mouth being widely open . . . by spitting, by blowing out of protruded lips; or by a sound of clearing the throat. . . . Extreme disgust is expressed by movements round the mouth identical to those preparatory to vomiting" (Darwin 1865, 257). Subsequent cross-cultural studies by Paul Ekman and colleagues (1972) and Eibl-Eibesfeldt (1989) have confirmed the universality, and by implication the innateness, of several human facial expressions of emotion. Van Hooff (1972;

1. Compressed-lips face
2. Play-face
3a. *Hoo* part of pant-hoot
3b. *Waa* part of pant-hoot
4a. Full open grin
4b. Full closed grin
5. Horizontal pout

FIGURE 7.1.
Facial expressions in chimpanzees. 1. Compressed-lips face is shown by aggressive chimpanzees; 2. Play-face is shown during play; 3a. & 3b. Facial expressions during different phases of the pant-hoot, a contact call given in a variety of contexts; 4a. & 4b. Variations of the full grin. Full open grin is shown by a very excited or frightened chimpanzee, the full closed grin is displayed by a chimpanzee who is less frightened or excited. The full closed grin is most similar to the human social smile (see text); 5. Horizontal pout is given in combination with whimpering, when a chimpanzee is frustrated.
Source: Line drawings from *In the Shadow of Man* by Jane Goodall. Copyright © 1971 by Hugo and Jane van Lawick-Goodall. Reprinted by permission of Houghton Mifflin Company. All rights reserved.

Van Hooff and Preuschoft 2003) has traced the evolution of the human smile from the silent bared teeth (appeasement) face, and the laugh from the relaxed open mouth (play) face of great apes.

Certain features of the human face, notably the eyes and the mouth, differ from those of chimpanzees. The eyes are accented by bright white sclera surrounding the iris, surrounded by long eyelashes, and shaded by eyebrows. Pink everted lips with two peaks on the upper lip frame the mouth. Their bright contrasting coloration and enlargement, combined with their use in facial expressions, suggest that they have been specialized for emotional displays. Use of makeup to exaggerate these features is consistent with this interpretation.

According to Desmond Morris (1967), "eye-white" or sclera surrounding the iris enhances eye signals. Morris points out that the eyes are signal detectors as well as signal transmitters. An increased area of white around the iris reveals where a person is looking. Since eyes with white sclera are "easier to read," Morris suggests that excessive "eye-white" signals sensitivity, weakness, and submissiveness (human females typically have more exposed sclera than human males). In contrast, reduced sclera obscure signals sent by the eyes, enhancing dominance. Signals emanating from the eyes are minimized by large brow ridges, which reduce the visible "eye-white." Morris suggests that thick hairy eyebrows serve as a "false contour," artificially enlarging the area, increasing the threatening nature of the display (see also Zahavi and Zahavi 1997). Human females typically reduce eyebrow hair because it appears more masculine and threatening. A thinner eyebrow provides a more juvenile appearance (Morris 1967).

Like the eyes, human lips are enlarged, contoured, and brightly colored. Like the eyes, lips are a focal point for social interaction for adults as well as infants. The shift from attacking by biting to attacking by hitting may have favored reduced canines to signal friendly approach via smiling. The mouth is the source of vocal communication, whether through emotional cries or words, as well as a key component of visual emotional displays. The mouth and oral cavity shape speech sounds in a manner that reveals the age and sex of the speaker. At puberty, female lips retain their childlike form, while male lips are surrounded by facial hair. Likewise, female voices retain their childlike features, while male voices deepen.

Juvenilization is the propensity of the adults of a species to retain youthful features of their ancestors (McKinney and McNamara 1991). In humans, juvenilized facial features include a baby face, higher voice, and a smooth complexion (Crook 1972). Guthrie (1976, 191) says the "everted red lip is more prominent among women who desire a baby-like appearance" because everted red lips are associated with nursing babies as they suck on the nipple. Retention of these traits in females suggests that selection favored juvenilization (Crook 1972). When males are constrained in their ability to obtain mates, they should prefer younger females because they have higher reproductive potential (Buss 1989; Small 1995). Thus, males may use juvenilization to assess female reproductive potential.

Bilateral Symmetry

Bilateral symmetry may signal overall condition. *Fluctuating asymmetry*, a lack of symmetry in traits that are symmetrical at a population level, may indicate developmental instability, that is, inability to resist the harmful effects of mutations, pathogens, or toxins during development (Scheib, Gangestad, et al. 1999). Indeed, fluctuating asymmetry in males is associated with reduced mating success, fewer sexual partners overall, later age at first copulation, and fewer sexual partners outside their primary relationship (Thornhill and Gangestad 1994; Gangestad and Thornhill 1997; Møller and Thornhill 1998).

According to the symmetry hypothesis, females use facial and/or body symmetry to assess the overall condition of potential mates, particularly whether a male has "good genes" (Scheib, Gangestad, et al. 1999). Thus symmetry translates into "attractiveness," so symmetrical faces are rated more attractive than asymmetrical ones (although some research suggests minor facial asymmetries may be more attractive than completely symmetrical faces: e.g., Swaddle and Cuthill 1995). If females cannot judge genetic quality of potential mates directly, they may use indirect physical cues like symmetry to do so. We suspect that males may also use this clue in their mate choice since average faces are considered most attractive (Galton 1869).

MALE DISPLAYS

Bipedal Displays

As humans became bipedal, males began fighting with their arms, with or without the aid of weapons, rather than their teeth. Consequently, shaking a raised fist is an effective threat display, with or without a weapon in hand. Both male competition and female choice have favored large, strong arm,

Desmond Morris's Naked Ape Scenario

Morris begins his book *The Naked Ape* (1967) by saying that a zoologist would study the human species by comparing us with closely related nonhuman primates. He speculates on the evolutionary history of our species, concluding that humans have "arisen essentially as primate predators" (21) that were able to blend their frugivorous heritage with meat-eating through bipedalism, increased brain size, the development of a fixed home base, cooperation and food-sharing, a sexual division of labor, pair-bonding, and male provisioning of females and young.

He notes that aside from differences in body structure related to locomotion, humans differ from other primates most notably in lack of visible body hair. We are not the least hairy primates, but because each hair is relatively short, we appear to have little body hair, and are functionally "naked." Morris summarizes various explanations for human "nakedness," concluding that the best explanation is related to overheating. He suggests that once our ancestors began hunting big game, they would have experienced significant overheating during long pursuits. Since hunting was crucial to their survival, there was intense selection pressure to reduce overheating, resulting in a "naked ape."

Morris begins his discussion of sex by comparing human sexual behavior with that of other living primates, and concludes that humans are "the sexiest primate alive" (1967, 53). The sexual behavior of humans, Morris suggests, evolved to enhance pair bonding, which was crucial to our ancestors because of increasing parental care demands. Continuous female receptivity, he suggests, encourages the continuation of the pair bond.

Morris discusses sexual signals, suggesting that upright posture has influenced our sexual signals. He says in our evolutionary past, females signaled their receptivity by displaying rounded buttocks and reddened vaginal lips, or labia, to the male (as other primates do today). With an upright posture, the breasts and red lips of the female mimic these features, and the male, already primed to respond when these signals were displayed from the genital region, would react similarly to these signals being displayed frontally. Morris also notes that our odor is a sexual signal. Humans secrete most of their scents through specialized glands, found primarily in the armpits and genitals, and Morris speculates that extra hair growth in those regions acts to trap these scents. He says the female orgasm is related to the pair bond, serving two purposes: (1) increasing the "reward" experienced by both partners, and (2) the chances of fertilization.

Morris also discusses cultural controls of sexual behavior, including the use of clothing to cover the genitals, making copulation a private act, reducing inadvertent touching of strangers, reducing body odors by bathing, and implementing sexual laws. Despite cultural restrictions on sexual behavior, cultures artificially exaggerate sexual signals through the use of bras and plastic surgery to enhance breast size and shape, the use of high heels to alter and enhance the swaying motion of the buttocks, cinching the waist to exaggerate the breasts and hips, use of lipstick to simulate the flushing of the lips during copulation, and the use of perfumes.

Morris begins his chapter on rearing by describing the processes of labor, birth, and nursing in humans. He explains that most mothers cradle their infants in their left arm because this brings the newborn close to the mother's heart, where it can be soothed by her heartbeat. Moving on from language development, Morris says crying is the first signal humans develop. He says laughing and smiling evolved from crying, as secondary signals, and speculates that the smile signals nonaggression, because it indicates fear combined with feelings of acceptance.

In his chapter on fighting, Morris says animals fight either to establish dominance in a social hierarchy, or to establish territorial rights over a piece of land, but humans, like some other animals, fight over both. Morris states that when humans began occupying home bases as pair-bonded families, aggression changed into group defense of shared territory, as well as defense of home base and family.

Morris discusses the physiological changes involved in aggression, including release of adrenaline into the bloodstream, increased respiration, and hair erection. These, along with other changes, prepare individuals for a fight by making energy immediately available. But, as Morris points out, full-blown aggressive attacks are relatively rare because they have been replaced by ritualized threats. He notes that many behaviors animals use in other circumstances have been elaborated into threat displays. These include defecation and urination, vocalizations, and hair erection, which make threatening display more obvious to opponents.

In the chapter on comfort, Morris discusses grooming in birds and mammals and the friendly behaviors that accompany it. Turning to humans, he suggests that, owing to lack of visible body hair, humans had to develop a different method of reinforcing friendly relationships. In place of grooming, Morris suggests that humans engage in "grooming talking" (1967, 167), or meaningless, polite conversation.

shoulder, and back muscles. Their high metabolic costs suggest that these are honest signals of male vigor and strength. Like other secondary sex characters, these organs enlarge at puberty under the influence of male sex hormones (Guthrie 1976). Aimed throwing of missiles is a universal form of aggression among adult humans, and aimed throwing games are universal among male children (Aldis 1975; Parker 1984). Cross-cultural studies of male aggression confirm the ubiquity of combat with weapons (Eibl-Eibesfeldt 1989). There is reason to believe that bipedalism evolved in part for aimed throwing of missiles at competitors and predators (see chapters 5 and 10).

Facial Hair

Guthrie (1976) points out that all primates other than humans attack members of their own species by biting them. Therefore intention movements like open-mouth tooth baring are threat signals in these species. Many nonhuman primates have chin or cheek tufts that create a false contour, making their faces appear bigger and more menacing. Likewise, Guthrie suggests, human beards make the jaw appear bigger and more menacing. He also suggests that larger, projecting chins and deeper voices evolved to enhance the threat display originating in the mouth. Visual displays are usually accompanied by verbal aggression, made more threatening by deeper voices of males (Guthrie 1976).

Guthrie (1976) notes that scalp hair in some species, like the gorilla, has the effect of increasing the height of the face, augmenting threatening facial signals. However, he notes, others use well-developed bare foreheads to increase the height of the face, increasing its potential as a threat signal. In these species, (e.g., the ukari), the color of the skin intensifies during threat displays, enhancing the signal (Guthrie 1976). According to Guthrie, changes in human facial color during threat displays or with age have a similar function and are more visible with the loss of scalp hair.

Male Genitalia

Although genitals are primary rather than secondary sexual characters, they are employed in courtship and copulation, and therefore have been subject to sexual selection (Eberhard 1985). According to Eberhard's (1985) "female choice" hypothesis, females discriminate among males based on their genitalia, and males with the preferred genitalia produce more offspring. He says the genitalia of some males may be better at entering the female, or placing sperm where they will fertilize the eggs, or may be more effective in eliciting essential female responses through stimulation (see the section on female orgasm). In other words, copulation is a continuation of courtship.

A key component of Eberhard's (1985) hypothesis is the arbitrary nature of the cues females use to discriminate among males. The ability of a male's genitalia to stimulate a female may have nothing to do with other aspects of his quality. This can lead to larger, fancier traits because females prefer ever more exaggerated male traits (supernormal stimuli) (West-Eberhard 1983). R. Fisher (1930) called this "runaway" sexual selection.

The human penis is larger than that of any other primate, both in proportion to body size and in absolute terms (Guthrie 1976; B. Smith 1984; Small 1995; figure 7.2b). Here we focus on two hypotheses for the human penis: the threat display hypothesis (Guthrie 1976), and ejaculate delivery hypothesis (B. Smith 1984).

Guthrie (1976) suggests that the large size of the human penis arose as part of a ritualized upright threat display (see also Small 1995). Following Westcott (1967), Jablonski and Chaplin (1993) suggest that habitual bipedalism evolved from stationary bipedal displays that resolved intragroup conflict through ritualized gestures. Guthrie (1976) and Morris (1967) both note that the upright posture of humans serves to expose the penis to public view, and that among males, penis size is associated with status, as well as female sexual satisfaction. In contrast, R. Smith (1984) suggests that a large penis was selected because it places sperm farther up the reproductive tract, closer to the egg, which increases the chances of fertilization, and also aids in sperm competition.

Testes Size and Sperm Competition

According to the "sperm competition" hypothesis, the sperm of different males compete to fertilize the egg *within* the reproductive tract of a female (R. Smith 1984). If females mate with two or more males around ovulation, males might succeed in fertilizing the egg by ejaculating more sperm. Since the amount of sperm individual spermatogenic cells produce is more or less constant across species, males should have larger testes, and thus, more spermatogenic cells, in species where females mate with many males (Harvey and Harcourt 1984).

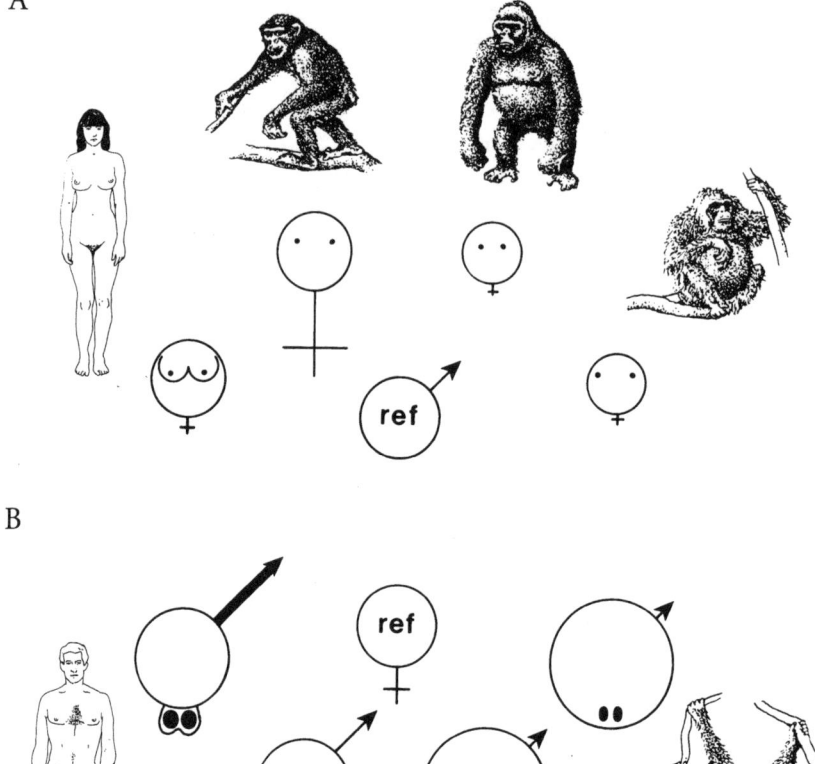

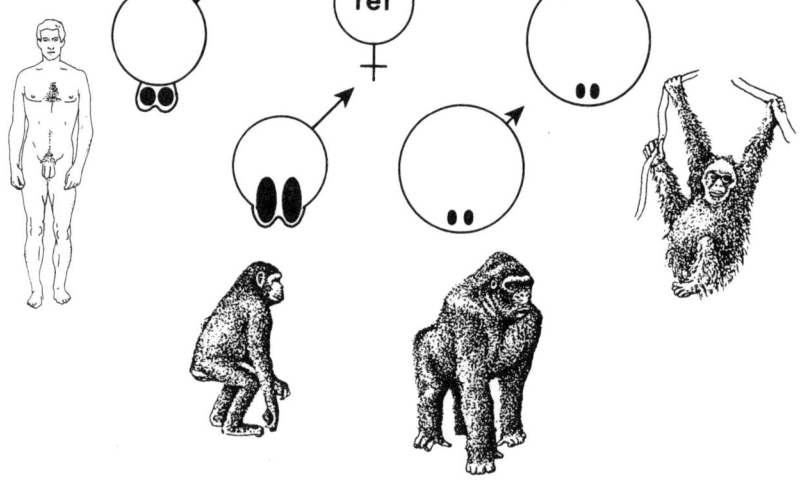

FIGURE 7.2.
Visual representation of sexual dimorphism in the Hominoidea. The size of the male and female symbols represents body size in relation to the body size of the opposite sex (labeled "ref"). In A, the relative size and position of female breasts are denoted by two dots within the circle, while sexual swelling is denoted by the elongated cross below the circle (only found in chimpanzees). In B, penis size is denoted by the length and thickness of the arrow attached to the circle while testes size and position are denoted by the two dots within (or below) the circle.

This material was originally published in "Human sperm competition," Robert L. Smith, in *Sperm Competition and the Evolution of Animal Mating Systems*, 601–59. Copyright © Academic Press 1984.

Comparing testes weight and body weight of primate species with different mating systems, Harvey and Harcourt (1984) found that *for a given body size*, genera in which females mate with many males at ovulation have larger testes than those in which females mate with only one male. In addition, they found that testes size translates into greater sperm production. In comparison with other great apes, human testes are relatively large, exceeded in size only by chimpanzee testes (R. Smith 1984; figure 7.2b). If testes size is a measure of sperm production, this indicates sperm competition has occurred in humans.

Whether enlarged testes are display organs in species with sperm competition is an interesting question we have not seen addressed. Two kinds of reports suggest that they may be. Chimpanzees often grasp the scrotum of males they are greeting (Nishida, Kano, et al. 2007). In a few reported cases, they have castrated males during fatal attacks (de Waal 1986; Nishida, Kano, et al. 1999). Both these behaviors are also seen in human males.

FEMALE DISPLAYS

Female Body Shape

As with males, human female displays were profoundly affected by the evolution of upright stance and locomotion. Unlike other primates, human females develop large, rounded breasts at puberty, which are retained throughout life (R. Smith 1984; figure 7.2a). Desmond Morris (1967) argues that the human female breast is primarily a sexual characteristic rather than a maternal one. The sexual signaling potential of human female breasts, he contends, is possible because they lack thick fur covering. Morris (1967) argues that at some time in our evolutionary past, females signaled their sexual receptivity by displaying rounded buttocks and reddened vaginal lips, or labia, to the male (as other primates do today). He suggests that the round breasts and red lips of the female mimic the buttocks and labia, and that males, already prepared to respond when these signals are displayed from the genital region, react similarly to these signals being displayed frontally. Thus, for Morris, female breasts signal female receptivity.

Researchers have shown that males in Western societies prefer relatively thin females with an hourglass figure, that is, large breasts and buttocks in combination with a narrow waist (e.g., Marlowe and Wetsman 2001; Jasieńska, Ziomkiewicz, et al. 2004; Marlowe, Apicella, et al. 2005). This can be quantified as a waist-to-hip ratio (WHR, or the circumference of the waist divided by the circumference of the hips) near 0.7 (Singh 1993).

In many mammalian species, males mate unselectively because they invest relatively little in offspring care (Trivers 1972). However, human males should be choosy because they invest heavily in offspring (Miller 2003). Specifically, they should prefer females who have high reproductive potential (i.e., are young and fertile). Whereas males can visually detect signs of age in females, they are unable to directly assess female fertility. The waist-to-hip ratio hypothesis is that females signal their reproductive potential (i.e., fertility) through physical cues.

High levels of estrogen at puberty result in fat deposition on the hips in females (reviewed in Singh 1993). Therefore, evolutionary biologists argue that the hourglass figure is an accurate indicator of reproductive status or potential in females (Singh 1993; Jasieńska, Ziomkiewicz, et al. 2004). Only females of reproductive age display the characteristic hourglass figure. Before puberty, boys and girls have similar (high) WHR, and after menopause, females' WHR again rises so that it is similar to male WHR (reviewed in Singh 1993). Thus, male WHR preference might signal young females with maximum reproductive potential who are not already pregnant (Marlowe, Apicella, et al. 2005).

Similarly, females with large breasts and small waists have higher reproductive hormone levels (Jasieńska, Ziomkiewicz, et al. 2004). Small breast size signals reduced probability of ovulation, either due to sexual immaturity or inadequate fat stores (Gallup 1982). Thus, according to the waist-to-hip ratio hypothesis, the hourglass figure is a signal of fertility and health in females, and males should prefer females with this body shape because such females should have higher reproductive potential. This is a good example of a sexually selected female trait via male choice (see table 7.1).

It should be noted that some human societies do not consider the hourglass figure the ideal female body shape (e.g., the Hadza: Marlowe and Wetsman 2001; Marlowe, Apicella, et al. 2005). In general, where food shortages are common or females' work is energetically expensive, males find heavier females (with higher WHR) more attractive because thinness probably indicates poor health (Marlowe and Wetsman 2001). Females suffering from poor health have lower reproductive potential than healthier females, so males who find heavier females (with higher WHR) attractive would have higher reproductive success (Marlowe and Wetsman 2001).

Female Orgasm

Whether female orgasms are displays is an interesting question (see Dixson 1998). Most attention has focused on whether female orgasm is an adaptation, or simply the byproduct of parallel male and female reproductive structures. According to Morris (1967), female orgasm played a critical role in the evolution of the pair bond. Like many anthropologists (see chapter 6), Morris (1967) notes that as brain size increased, development slowed, and offspring were dependent for longer periods of time. He argues that paternal behavior developed through a strong pair bond, which kept males and females attached while they reared their slow-developing offspring.

The sperm manipulation hypothesis is that orgasm allows females to manipulate sperm retention inside their reproductive tracts. Changes in the uterus following orgasm, specifically a drop in pressure, create suction, which may help pull semen into the reproductive tract, thereby promoting fertilization. Accordingly, orgasms experienced between copulations reduce sperm retention during the next copulation (Baker and Bellis 1993).

Donald Symons (1979) first proposed that female orgasm is nonadaptive. According to Symons, "human female orgasm is best regarded as a potential" which is present in all mammals, but expresses itself in humans due to "sufficiently intense and uninterrupted stimulation" (89). He says this "potential" is a by-product of the common embryonic source of the tissues involved in orgasm in males and females.

Lack of Estrus/Concealed Ovulation

Much has been written regarding concealed ovulation and lack of estrus in human females. The term *estrus* refers to "a relatively brief period of proceptivity, receptivity, and attractivity in female mammals" which usually coincides with ovulation (Symons 1979, 97). *Concealed ovulation* refers to lack of visible signs of ovulation (e.g., a sexual swelling) in humans (Burley 1979; Symons 1979) and some other primates.

Lack of estrus means that the human female can *potentially* engage in copulation at any time during the menstrual cycle. Two functional hypotheses have been proposed for this phenomenon: (1) concealed ovulation evolved to enhance pair bonding via frequent copulation outside of ovulation (e.g., Morris 1967; Alexander and Noonan 1979; Symons 1979; Lovejoy 1981), and (2) concealed ovulation confuses, rather than clarifies, paternity (Hrdy 1979, 1981).

Morris (1967) notes that human females ovulate only once during the menstrual cycle, but have the potential to copulate throughout. Therefore, the majority of human copulations must not be concerned with reproduction (see also Lovejoy 1981). According to Morris, the human female's ability to copulate throughout the reproductive cycle, during pregnancy, and shortly after giving birth, safeguards the pair bond by reducing the probability that her mate is sexually frustrated.

Extended periods of female sexual receptivity combined with males' inability to detect ovulation would have reduced their confidence in paternity of their mates' offspring (Symons 1979). To increase paternity confidence once estrus was lost, males would have to increase the time and energy devoted to copulating with and guarding their mates (Symons 1979; Lovejoy 1981). This would increase confidence of paternity, resulting in higher levels of paternal care of the offspring (Alexander and Noonan 1979).

Sarah Blaffer Hrdy (1979, 1981) suggests mating with many males allows females to reduce the chances that males will kill their offspring. If females display no outward signs of estrus, males will be less likely to gauge when ovulation occurs. In species where infanticide occurs, males use a rule of thumb to determine whether they are the fathers of newborn infants: "if I mated with the infant's mother before she gave birth, I may be the father." Such a rule may keep males from killing other males' offspring, but will prevent them from killing their own.

THE EVOLUTION OF HUMAN BODILY DISPLAYS

As with other characters, we must rely on comparative primate data to reconstruct traits of the last common ancestor (LCA). In contrast to other characters, however, ancestral bodily displays are difficult to reconstruct because skin and hair do not readily fossilize (see also Miller 2003). Whenever possible, we try to identify when bodily displays may have appeared in a stage of hominin evolution and why. Bipedalism is the most reliable marker for some displays. Degrees of sexual dimorphism may offer additional clues to mating systems.

Bodily Displays in Chimpanzees and the LCA

Using the chimpanzee as a model for the LCA may give us some clues to the evolution of human displays. Chimpanzees and bonobos are polygamous/promiscuous maters. Chimpanzee testes are five and ten times larger than the orangutan's and the gorilla's, respectively, and the penis is more than twice as long as that of the gorilla (R. Smith 1984; figure 7.2b). The result is that a male chimpanzee can ejaculate more sperm than either the gorilla or orangutan (Short 1981). In addition to sperm competition, male chimpanzees compete by monopolizing females during consortships (Goodall 1986).

Female chimpanzees signal sexual receptivity (estrus) with a sexual swelling (figure 7.2a) and prefer to mate promiscuously. The presence of sexual swellings on a regular cycle indicates that a female chimpanzee is of reproductive age (sexual swellings begin at menarche), and each sexual swelling itself is a signal to males that ovulation is imminent. Although a sexual swelling can last for several weeks, maximum swelling indicates that ovulation is near, and matings peak during this time (M. Thompson 2005).

Loss of estrus scenarios are based on the assumption that, like chimpanzees, the LCA had estrus (Pawlowski 1999). However, recent research suggests that concealed ovulation is more widespread in primates than previously appreciated (e.g., hanuman langurs and vervet monkeys) (Hrdy 1977; Andelman 1987; Heistermann, Ziegler, et al. 2001). Indeed, great apes other than chimpanzees and bonobos, as well as lesser apes, lack estrus displays. So logically, estrus displays could have evolved either before or after chimpanzees and bonobos diverged from hominins.

Phylogenetic studies indicate that sexual swellings are most common among polygamous species, and were never present in the hominin line (Sillén-Tullberg and Møller 1993; Dixson 1998). Rather than "concealed" ovulation evolving on the hominin line, *sexual swelling* may have evolved on the line leading to chimpanzees and bonobos (Campbell 2007; see figure 7.2a). It is important to remember, however, that sexual selection leading to speciation can change sexual behavior rapidly (West-Eberhard 1983).

Like loss of estrus scenarios, female orgasm scenarios are based on the assumption that the trait is absent in great apes, particularly chimpanzees and bonobos, and hence in the LCA. Dixson's (1998) review indicates widespread occurrence of female orgasm (specific vocalizations and visual signals associated with vaginal and uterine contractions) among female primates, particularly those in polygamous mating systems. He concludes, "the capacity to exhibit orgasm in the human female is an inheritance from ape-like ancestors" (Dixson 1998, 133).

These comparative studies suggest that we should avoid devising scenarios for human traits until phylogenetic studies demonstrate that these traits are uniquely derived in our lineage (rather than shared derived characters also present in great apes and possibly other primates; see chapter 2). In retrospect, we can see that such studies would have eliminated hominin scenarios dealing with loss of estrus and female orgasm. Given these new insights, we have dropped these scenarios from further consideration.

Bodily Displays in Middle-Period Human Ancestors

Bipedalism and loss of body hair have altered many of the signals displayed by the LCA. *Homo erectus/ergaster* were apparently the first hominins fully committed to bipedal stance and locomotion (see chapter 5). This suggests that they had begun to evolve some of the bodily displays associated with upright stance. These would include heavily muscled arms, backs, and shoulders and enlarged penises in males, and enlarged breasts and hips in females. Nakedness, which would have enhanced the visibility of these displays, could have evolved in response to the heat load of bipedal running in the savanna (Wheeler 1984, 1991, 1992), and/or after our ancestors devised shelter and clothing to protect their skin from extreme climates (Pagel and Bodmer 2005).

Homo erectus/ergaster were the first hominins found outside of tropical, equatorial Africa. They may have controlled use of fire. Their technological innovations suggest the cognitive ability to make devices to protect them from the weather (see chapter 8). Once clothing, shelter, and fire were established, selection pressure for a thick layer of body hair would have been relaxed. In addition, selection would have favored reduced body hair as a signal of good health reflecting reduced parasite loads. We agree with Darwin that female competition and male choice would have favored greater hair loss in females compared to males as females competed for male parental investment.

As infants lost their ability to cling attendant on committed bipedalism, distal vocal and visual signals would have augmented tactile communication. Likewise, increased kin and parental investment in dependent offspring would have favored more distance communication displays between infants and caretakers. Both white eyes and pink lips would have made visual displays more compelling, helping infants and children track the attention of their caretakers and helping caretakers track the attention of the youngsters (Parker

2000). Likewise, vocal signals would have become more important as foraging caretakers were out of sight of offspring (Falk 2004; see chapter 9). Subsequently, bonding signals of smiling and laughing and kissing between mother and offspring might have become incorporated into courtship displays (Eibl-Eibesfeldt 1989).

Most anthropologists believe pair bonds arose in conjunction with increased hunting and scavenging, increased meat consumption, and increased brain size. They expect these factors converge in greater parental and or kin investment in dependent offspring. Gauging the appearance of pair bonds is difficult. Sexual dimorphism of fossil species is one clue to their mating systems. *Homo erectus/ergaster* have been viewed as good candidates for the originator of pair bonds because of their apparently low degree of dimorphism (McHenry 1996), but a recent discovery of a very small female skull suggests that this species may have been highly dimorphic. This suggests that they had one-male harems (Spoor, Leakey, et al. 2007). For this reason we question the common view that pair bonding arose in this species.

Bodily Displays in Modern Humans

Apparent discontinuity in the evolution of bodily displays arises first from the absence of fossil data, which leaves the comparison between ourselves and African apes, and second from sexual selection's role in creating new mating displays during speciation. Whatever their history, these bodily displays were present in anatomically modern humans. From the Middle Stone Age on, our ancestors have elaborated these displays. Most notably, they have elaborated male and female signals, as well as facial displays through hair styles, clothing, makeup, and jewelry. For example, archaeological evidence suggests that humans have been using ornaments and pigments to enhance their appearance at least since the Middle Stone Age (R. White 1989; McBrearty and Brooks 2000). Historical and cross-cultural studies also reveal increased ingenuity and preoccupation with status.

CONCLUSIONS ON CULTURE, BEAUTY, AND THE EXAGGERATION OF BODILY DISPLAYS

All human societies have standards of beauty and propriety. In *The Descent of Man*, Charles Darwin (1871) discusses differing ideas of beauty, including preferences for particular face and body shapes, skin colors, differing degrees and styles of head and facial hair, and body and facial ornamentation, such as tattooing, scarification, and piercing. In all human societies, many members emulate the "ideal" standard of beauty. Their purpose is to attract members of the opposite sex and also to display their status and group membership.

Far from obscuring bodily displays, clothing and ornaments *augment* and *exaggerate* sexual signals (Morris 1967). Although females in Western societies cover their breasts, brassieres shape and often enhance the breasts. Necklaces, earrings, and bracelets also accent the breasts and face. High-heel shoes distort females' gait and increase swaying of hips, drawing attention to the hips and buttocks. Tight belts or corsets pull in the waist to exaggerate the hips and the breasts, increasing the hourglass figure stimulus. Likewise, males' clothes accentuate height and broad shoulders, increasing the potential threat display. Even more ornate exaggerations of bodily displays were popular during the Victorian (wearing bustles to artificially enlarge the female buttocks) and Renaissance (wearing codpieces to artificially enlarge the penis) eras. Likewise, ornate styles augment and elaborate the size and shape of head and facial hair, fingernails, and feet (Bell 1977). Indeed, Thierry (2005) suggests that both reduced body hair and long head hair evolved as culturally mediated cues to tribal membership, rank, age, and marital status.

Dark eyeliner and eye shadow offset the sclera or "eye-white," exaggerating the sensitivity and submissive signals that are projected (Guthrie 1976). In much the same way, wearing red lipstick and blush heightens sexual signals (Morris 1967). Rouge simulates the flush experienced during sexual arousal. If adult males are selected to cue in on these signals during copulation, they should arouse their interest.

We tend to think of these commonplace aspects of dress and beauty in Western society as being driven solely by culture. However, an evolutionary perspective allows us to see that in all of these cases, the cultural preferences for specific makeup and clothing styles and/or ornaments serve to exaggerate or augment sexual signals and bodily displays. In other words, there is a biological basis for our culturally defined views of beauty and attractiveness. In Robin Fox's words, we are "cultural animals" (Fox 1971).

8

Origins of Language

> I cannot doubt that language owes its origin to the imitation and modification of various sounds, the voices of other animals, and man's own instinctive cries, added to signs and gestures. When we treat of sexual selection, we shall see that primeval man, or rather some early progenitor of man, probably first used his voice in producing true musical cadences, that is, singing . . . and we may conclude from a widely spread analogy, that this power would have been especially exerted during the courtship of the sexes.
>
> —*Darwin 1871, 463*

Long considered the defining characteristic distinguishing humans from other animals, language has fascinated anthropologists. So great was the impulse to explain its origins and so scant the evidence for explanatory scenarios that in 1866 the Linguistic Society of Paris announced a moratorium on speculations on language evolution. This moratorium was officially breached in 1976 with the publication of an entire issue devoted to this topic in the *Annals of the New York Academy of Sciences* (volume 280; Gibson and Ingold 1993).

The difficulty in reconstructing language evolution arises both from the complexity and the uniqueness of language. There are no language fossils, no other language-producing species, and no surviving protolanguages. Language has many dimensions. First, many bodily organs normally interact in the production and comprehension of spoken language. Clearly, the human mouth, teeth, tongue, vocal tract, and auditory system, as well as the motor and perceptual apparatus of the brain, have been specialized for speech production and comprehension. Likewise, the diaphragm, the intercostal muscles, and the spinal nerves serving thoracic elements involved in breathing have been specialized for control of the breath in speech.

Second, the speed, complexity, and efficiency of spoken language depend upon automated production and comprehension of speech sounds organized into syllables, words, and sentences capable of encoding an infinite variety of meanings. Among other things, grammars encode temporal, spatial, and causal relations of actors, objects, locations, instruments, and patients (or recipients). Language develops in several stages commencing at birth. The cognitive and social complexity of language is such that children take at least fifteen years to master it fully. The heart of language, whether expressed in speech, hand signs, or writing, is a symbolic system encoding meanings.

When did language arise? What evolutionary stages or sequences of emergence occurred? What would a primitive protolanguage look like? Was language always vocal, or was it once gestural, or was it always both? What selection pressures shaped it? What kinds of clues can we use to discover its history?

When addressing these questions, we should be clear about whether we are discussing the evolution of proximate mechanisms of language production and comprehension (the vocal tract or the brain, etc.) or whether we are discussing the adaptive value of language and how it may have been shaped by selection. In addition, we have the ongoing challenge of relating language to culture and cognition.

The best clues to the earliest protolanguages come from four sources: ape language studies, studies of language development in deaf and hearing children reared in different environments, studies of the invention of creoles from pidgins, and the invention of sign languages (Pinker 1994).

One clue to innate elements of language has come from the discovery of universal stages of language acquisition from birth to three years of life, and their correspondence with stages of cognitive development (R. Brown 1973; Bates 1976; Nelson 1996). Another clue has come from the discovery that deaf children of hearing parents spontaneously develop symbolic signs (Goldin-Meadow and Mylander 1998). Other clues have come from the study of the spontaneous origins of creole languages from pidgins (Bickerton 1981), and the spontaneous origins of spatial grammar in Nicaraguan sign language (Senghas, Kita, et al. 2004). Before discussing various evolutionary scenarios, it is important to reconstruct the communication patterns of our last common ancestor with chimpanzees and bonobos.

COMMUNICATION CAPACITIES OF HUMAN ANCESTORS
Communication of Last Common Ancestor (LCA)

Since neither modern African apes nor more distantly related Asian apes display morphological and neurological capabilities for speech, we can assume that the last common ancestor also lacked these adaptations. For clues as to how these creatures communicated, we look to our closest living relatives: Chimpanzees and other great apes communicate through postures, gestures, facial expressions, and vocalizations. Like Old World monkeys, lesser apes, and orangutans, African apes have hairless faces and highly mobile facial muscles and lips, which highlight their expressive facial movements. Facial expressions of emotion include grins of fear and excitement, compressed lips of anger, pouts of distress, relaxed open-mouth face with upper teeth covered, and associated vocalizations (e.g., Goodall 1986; see also chapter 7).

In other words, chimpanzees and other African apes express their intentions (aggressive, submissive, or friendly) through a variety of communicative sounds, postures, and gestures, some of which are similar to those of modern humans (Goodall 1968; Goodall 1986). Chimpanzees, bonobos, gorillas, and orangutans display vocal, manual, postural, and facial communication systems that are primarily emotional rather than referential. They do, however, have some voluntary control over these actions as seen, for example, in their attempts at deception (de Waal 1983).

In addition, great apes display symbolic abilities similar to those of two-year-old human children (R. Brown 1973; Savage-Rumbaugh, Murphy, et al. 1993). These include production of symbolic gestures and combinations of these gestures encoding such relationships as agent-action, action-object, agent-object, goal-action, action-goal, location-entity, etc. (Greenfield and Savage-Rumbaugh 1990). It has been known for more than a quarter of a century that captive great apes are able to learn and use more than three hundred arbitrary symbols to communicate with humans and with each other. They can ask who, what, and where questions (Gardner, Gardner, et al. 1989).

Hockett and Ascher's Revolutionary Language Origins Scenario

Charles Hockett and Robert Ascher (1964) tell a story of the emergence of the first humans from their protohuman ancestors, emphasizing the evolution of language. They characterize this as a revolution, that is, as an irreversible process.

Following what they characterize as the majority opinion in anthropology, Hockett and Ascher trace the human lineage back to its "point of separation from the ancestors of the great apes, the gibbons, and the siamangs" (1964, 137). Based on the dryopithecine fossil *Proconsul*, they infer that protohominoids (now protohominids) lived in East Africa in the Lower Miocene or Upper Oligocene. They suggest that these creatures were expert climbers, which may have brachiated, and could walk on all fours and stand semierect. They may have picked up sticks and stones to use as tools. Their life history was like that of lesser apes. They displayed little sexual dimorphism, were tailless, and had large canine teeth. They ate a largely vegetarian diet supplemented by occasional small prey. They lived in territorial bands of ten to thirty individuals, slept in nests, and used a variety of communicative signals including vocalizations. They propose that these protohominids had a call system similar to that of gibbons.

They describe a call system as including "a repertory of half a dozen distinct signals, each the appropriate vocal response—or the vocal segment of a more inclusive response—to a recurrent and biologically important situation" (Hockett and Ascher 1964, 139). These signals are mutually exclusive and cannot be combined or mixed: "The technical description of this mutual exclusiveness is to say that the system is *closed*. Language in sharp contrast, is *open or productive:* we freely emit utterances that we have never said nor heard before" (139). They say that a call system also lacks other crucial features of language including *displacement*, that is, the ability to communicate about things out of sight or time, and *duality of patterning*, that is, combinations of meaningless phonological elements into meaningful elements. It also lacks transmission of traditional forms through social learning.

Hockett and Ascher suggest that our ancestors were pushed out of the trees by competing apes when a climatic change in the Miocene reduced forest cover. They survived by the trick of carrying infants and brush for nests as well as sticks and stone tools for scavenging and predation. Hunting stimulated collaboration and visual attention to quarry and other participants: "Collective hunting, general food-sharing, and the carrying of an increasing variety of things all press towards a more complex social organization, which is only possible with more flexible communication" (1964, 142).

The authors go on to propose a scenario for the transition from a closed to an open call system, the prerequisite for the emergence of language. They suggest that instead of giving a clear food call or a clear danger call, a protohominid (now protohominin) with a larger but still closed repertoire uttered a blended call with some of the characteristics of each of these. They further suggest that other members of the group responded appropriately: "Thus reinforced, the habit of blending two old calls to produce a new one would gain ground" (Hockett and Ascher 1964, 142). They imagine that ten calls could blend creating a hundred calls, "each of the possible calls has 2 parts, and each part recurs in other whole calls. One has the basis for *building* composite signals out of meaningful parts, whether or not those parts occur alone as whole signals" (143). They call the minimal meaningful signal elements "premorphemes." Over thousands of years, an open call system or *prelanguage* was achieved, which could only be transmitted through teaching and learning. This prelanguage, however, lacked duality of patterning.

According to their scenario, as the stock of minimal meaningful signals increased, the acoustical/articulatory space for their perception and production became increasingly packed, and pressures on the system favored a shift from a premorphemic toward a morphemic system: "Pre-morphemes began to be listened to and identified not in terms of their acoustic gestalts but in terms of smaller features of sound that occurred in them in varying arrangements. In pace with this . . . articulatory motions came to be directed towards the sufficiently precise production of the relevant smaller features of sound that identify one pre-morpheme as over against others" (Hockett and Ascher 1964, 144). Combinations of these smaller features of sound, the phonemes, came to be recognized as identifying different morphemes: "The phonological system of a language has almost as its sole function that of keeping meaningful utterances apart" (145).

Hockett and Ascher argue that language had emerged from prelanguage by about one million years ago. They argue that the enlarged brain could only have evolved if humans had already achieved the "essence of language and culture" (1964, 145). They further argue that duality of patterning arose at the same time. They say that the reduction in tooth and jaw size plus the habit of talking favored "precise motions of lips, jaw, tongue, velum, glottis, and pulmonary musculature," leading to the evolution of the organs of speech in fossils (146). In addition, they say that the uniformity in certain features of language suggests that certain critical features of language developed before the diaspora of humans out of Africa. The authors revised parts of their scenario in response to the many commentaries published with their article.

Gordon Hewes's Language Evolution Scenario

Gordon Hewes (1973) was one of the first anthropologists to look to ape language studies for clues to language evolution. He notes that whereas most theories of language origin assume a vocal origin, the idea of the gestural origin also has a long history, going back at least to the Enlightenment, and was espoused by some early anthropologists, for example, Tyler and Morgan, long before these studies.

Hewes frames his argument along the following lines: First, he argues that protolanguages must have been simpler and more restrictive than any existing spoken language. Second, he argues that australopithecines must have had at least the cognitive capacities of living African apes. He also notes that the emotion-driven vocalizations of living primates lack the voluntary control of manual behaviors of great apes. Citing work by Emile Menzel, he reviews various chimpanzee behaviors which could be a substrate for gestural language, including use and decoding of gestures, postures, and locomotion, to exchange environmental information. He notes that "modern pongids, and hence almost certainly early hominids, can do more than decode such signals: they can *imitate* them" (Hewes 1973, 8). Therefore, he concludes that early hominids (now hominins) would have been able to acquire simple sign languages similar to those inculcated in chimpanzees. He notes, however, these could not have resembled modern sign languages, which are influenced by speech and writing: "The semantic productivity of various finger, hand, arm, head, and facial gestures, all within the anatomical and intellectual capabilities of the australopithecines is impressive, even though we have no way of proving that it was used" (Hewes 1973, 9).

Hewes's reconstruction of early hominid gestural protolanguage includes imitation of characteristic animal behaviors as signs for animals and imitation of characteristic tool use operations as signs for the tools. He also notes that bodily functions of eating and drinking, etc., are easily mimed.

Moreover, they can refer to objects and events outside their perceptual field. Their comprehension of simple sentences is comparable to that of a two-year-old child (Savage-Rumbaugh, Murphy, et al. 1993). In contrast to children, who achieve similar milestones in the first two years of life, great apes only reach their peak cognitive achievements at eight or nine years of age.

Interestingly, whereas great apes can learn manual or visual signs (e.g., pictographs or computer keys), they cannot utter words. Their capacity for sign-

He argues that cortical lateralization preceded speech and arose from selection for separation of power from precision grip, and for right-left consistency in reference to landmarks.

Hewes notes several advantages of vocal over gestural language, including its lower energy cost and its free availability as compared to the gestural-visual channel, which is continually faced with information from nonlanguage sources. Admitting that the shift from gestural to vocal language involved significant changes in the lips, tongue, larynx, and brain, Hewes suggests that the long evolutionary history of hunting played a crucial role in preparing hominids for decoding speech by "selecting for cross-modal cognitive analysis of a wide range of environmental noises, from the cries and calls of different species to the sounds made by snapping twigs or rustling underbrush" (1973, 9). He describes the "mouth-gesture hypothesis" that vocal sounds could have come to be systematically linked with elements of manual-gestural signs through mouth, tongue, and lip imitation of gestural movements. He appeals to evidence for some semantic-phonetic universals in sound symbolism: "High front vowels are associated with smallness, whereas low or back vowels are indicative of flatness or large size. Sharp pointed things are more likely to be associated with t or k sounds, while soft, smooth things tend to be associated with l or m sounds" (10). However speech arose, he notes, most linguists argue that modern spoken language arose in the Upper Paleolithic.

Finally, Hewes notes that gestural symbols have persisted as paralinguistic accompaniment to speech and were transformed into "frozen gestures" of drawing, painting, and sculpture. Moreover, he argues that "the old visual-gestural channel became the preferred mode for advanced propositional communication in higher mathematics, physics, chemistry, biology and other sciences and technology" (1973, 11). He explicitly assumes that cultural evolution has been the prime mover for language evolution.

ing corresponds to their greater ability to imitate manual actions as opposed to vocalizations. Great apes use symbols productively to classify objects into categories of tools and food, colors and shapes, and to request objects and activities that are not present. They can learn symbols without training, and teach them to conspecifics (Fouts, Fouts, et al. 1984, 1989). The symbolic abilities of chimpanzees, bonobos, gorillas, and orangutans are all similar to those of two-year-old human infants (Gardner and Gardner 1969; Savage-Rumbaugh,

Rumbaugh, et al. 1978; Patterson 1980; Gardner, Gardner, et al. 1989; Savage-Rumbaugh, Romski, et al. 1989; Miles 1990; Bonvillian and Patterson 1993; Savage-Rumbaugh 1995; Jensvold and Gardner 2000). It is important to note in this context that chimpanzees display enlargement of the left temporal area (the planum temporale in Wernicke's area) of the brain associated with language comprehension in humans (Gannon, Holloway, et al. 1998).

Several investigators have cast doubt on the symbolic abilities of great apes (Terrace, Petitito, et al. 1979), claiming that they are simply conditioned reactions. The evidence does not support this claim. First, the demonstration of similar linguistic abilities of all the great apes but no other nonhuman primates argues against this claim. Second, these abilities are consistent with (and even less developed than) other mental abilities of great apes, notably their logical and physical cognition (Chevalier-Skolnikoff 1976; Langer 1993, 1996; Parker and McKinney 1999). Third, a comparative study of spontaneous interaction versus conditioning methods revealed that the same individual (Terrace's chimpanzee subject, Nim) displayed greater symbolic ability in a spontaneous interactive setting than he did in a conditioning mode (O'Sullivan and Yeager 1989).

A key question is whether these symbolic abilities are expressed under natural conditions. One clue comes from a study of captive bonobos showing that they developed signs for requesting copulatory positions (Savage-Rumbaugh, Wilkerson, et al. 1977). Likewise, recent work indicates that gorillas are capable of spontaneously developing symbolic gestures through conventionalization of social interactions (J. Tanner and Byrne 1996, 1999). In other words, their symbolic ability is expressed without human tutelage. One possible case of symbolic communication has been described in wild Taï chimpanzees (Boesch 1991a). Given the symbolic capacities of living great apes we surmise that the LCA of humans and chimpanzees must have displayed at least a great ape level of symbolic ability, and perhaps a higher level.

Anthropologists have expended much effort speculating on when the capability for modern human language arose. Direct evidence of language is nonexistent in the archaeological record, except in the form of writing beginning six thousand years ago, long after modern humans appeared. The consensus is that spoken language has been in use much longer, at least since the appearance of anatomically modern humans between one hundred and sixty thousand and sixty thousand years ago (Donald 1991; Mithen 1993; Noble and Davidson 1996). The earliest language reconstruction backward from living languages goes back a maximum of ten thousand to twelve thousand years ago (Nichols 1998).

In analyzing fossils, paleoanthropologists have looked to four kinds of anatomical data to discover which hominin species displayed the ability to use spoken language: (1) evidence of language-related brain structures in endocasts of fossil skulls, (2) brain size and organization, (3) evidence of humanlike vocal tract structures in the base of the skull and face, and (4) evidence of increased vocal control based on spinal cord size and chest muscles.

Communication among Early Human Ancestors

The brain sizes of australopithecines are only marginally larger than those of chimpanzees. In terms of brain structure, some "language centers" (e.g., Broca's area for speech production and Wernicke's area for speech comprehension) have been identified. Based on the presence of Broca's area in early *Homo* (but not in *Australopithecus*) endocasts, some researchers believe that some language capabilities were present in *Homo* but not earlier hominins (Tobias 1987). This claim needs to be reconciled with data suggesting that in modern humans, these structures differ among speakers of different languages (Elman, Bates, et al. 1996). Likewise, fossilized traces of patterns of blood flow involved in cooling the brain reveal that the human pattern evolved after the australopithecines (Falk 1992).

Previous studies of the base of the skull indicated that australopithecines probably had straight (single tube) vocal tracts unlike the right-angled (two-tube) vocal tracts of modern humans, similar to those of monkeys and apes. More recent work, however, indicates that measurements of cranial base angulation cannot be used to reconstruct the evolution of the vocal tract in apes (Liebermann and McCarthy 1999). The authors of this study note that spatial relations among the jaw, the palate, and the hyoid bone (a free-floating bone that anchors the tongue inside the jaw) give the vertical projection of the vocal tract, and they could be useful in analyzing fossils. Based on the hyoid bone of Neanderthals, some paleoanthropologists suggest that this species had spoken language (Arensburg, Tullier, et al. 1989).

Armstrong, Stokoe, et al.'s Origins of Syntax Scenario

Armstrong, Stokoe, et al. (1994) present a gradualist scenario for the gestural origins of syntax based on differences in gestural and spoken language. The authors argue that the greater iconicity and the phraselike structure of signs provide clues to the origins of grammar, the most difficult element of language evolution to explain. They emphasize that both spoken and signed languages can be described in terms of neuromuscular activity or gesture, noting, however, that much more work has been done in analyzing spoken language in these terms.

The authors note that one of the defining characteristics of language is duality of patterning. According to this linguistic concept, each language uses a small number of speech sounds or movements, which are combined into the smallest meaningful units (e.g., -ing, -s): "Duality, thus, is between the meaningless and the meaningful," that is, between the phonological and the semantic levels (Armstrong, Stokoe, et al. 1994, 352). They note that problems arise in trying to show the separation of these levels in sign language. According to Stokoe's semantic phonology, "a word (sign) of a primary sign language may be seen as a marriage of a *gestural noun* and a *gestural verb*. . . . The sign itself may be construed as an agent-verb construction. The agent is so called because it is what acts . . . and the verb is what the agent does" (352). Thus, they claim that gestures collapse the symbolic link between semantic and phonological levels.

The significance of this is that visible gestures reveal the rudiments of syntax in the parts of a gesture such as miming throwing or catching. The iconicity of these signs resides in their whole "'gestalt' of something acting, its action, and the result, the patient it acts upon" (Armstrong, Stokoe, et al. 1994, 355). Such gestures can be read as either words or

---

Another possible source of information about early hominin language abilities is the complexity of material culture remains. The material culture remains of basal hominins are nonexistent, and those of early hominins, the australopithecines, are meager. The most important archaeological remains of early hominin ancestors are the cobblestones and cracked marrow bones associated with *A. garhi* in Ethiopia (Asfaw, White, et al. 1999; see chapters 4 and 10).

Communication of Middle-Period Hominins

Despite the essentially modern bipedalism of *H. erectus*, the spinal cord and thorax of this species suggests that their neurological control of breathing was insufficient for producing modern human speech (MacLarnon and Hewitt

sentences, depending on special added movements of the face, eye, or arm. These often ignored aspects are clues to the origin of syntax: "In this scenario, getting to syntax was a matter of analysis rather than synthesis" (353). This analytic approach contrasts with the more usual synthetic approach of imagining the combination of spoken units into a grammar. It also provides continuity with abilities present in the common ancestor of chimpanzees and humans. They note that chimpanzees show the ability to use both symbols and icons, the prerequisites for language evolution, consistent with the continuity approach. They also note the proposal that a cross-modal association area in the inferior parietal lobe, present in chimpanzees, is vital for symbolizing.

Based on evidence that bipedalism freed the hands and that the human hand evolved its modern form long before the vocal tract, the authors argue that gestural communication would have led, but that vocal communication would have been important from the beginning of hominid (now hominin) evolution: "The scenario we propose involves an evolutionary state in which visible gestures, for all the reasons proposed above, take the lead with respect to flexibility of output and critically, the elaboration of syntax. The cognitive apparatus necessary to support these developments would then have been available for elaboration in spoken language as well" (Armstrong, Stokoe, et al. 1994, 358).

Their scenario focuses on deriving syntax from presyntactic behavior rather than on selective pressures or the timing of events. Nevertheless, they quote Tobias to the effect that endocasts of the brains of *Homo habilis* are the first to display neurological prerequisites for language, a strong inferior parietal lobe and a prominent Broca's area, which they suggest supports signing as well as speech.

1999). They displayed larger brain size (about 900 cc, or two-thirds the size of modern human brains), but apparently had life histories more like their early hominin ancestors than those of modern humans (Dean, Leakey, et al. 2001). Although there is no magic threshold of brain size that denotes capacity for language, larger brains imply greater information-processing abilities, and therefore greater likelihood of language (Gibson 1988). *Homo erectus* also displayed bifaced Acheulean tools, construction of shelters, scavenging or hunting large game, and probable use of fire as they moved out of Africa. All these factors suggest that *H. erectus* displayed increasing social coordination and some form of referential communication, allowing more prolonged apprenticeships in tool production and use and more foraging than their ancestors.

Communication among Recent Human Ancestors

Early archaic *H. sapiens* had reached Europe by about eight hundred thousand years ago (Carbonell, Bermudez de Castro, et al. 1995; Bermudez de Castro, Arsuaga, et al. 1997). Later archaic *H. sapiens* date from about two hundred and fifty thousand years ago (Conroy 1997). They had larger brains than their ancestors, but more robust skulls and teeth than anatomically modern humans. The latter group also had more complex and varied material culture than their Middle Period ancestors. The best known of these are the Middle Stone Age cultures of Africa. They are similar to the Mousterian and the late-developing Chatelperronian of the Neanderthals in Europe. Archaeologists disagree over whether Chatelperronian should be classified as Mousterian or Aurignacian (Harrold 1989; McBrearty and Brooks 2000).

Anatomically modern *H. sapiens*, who migrated out of Africa about sixty thousand years ago into the Middle East, Europe, Asia, and Australia, had still more complex and varied material cultures. The best known of these are the Upper Paleolithic cultures in Europe dating from about forty-five thousand to thirty-five thousand years ago. The Aurignacian was the first of a rapidly changing series of regionally specialized cultures of these modern *H. sapiens* that succeeded each other in the period dubbed the Upper Paleolithic Revolution.

The Upper Paleolithic Revolution was distinguished from earlier Middle Stone Age cultures by many features. These include more blade and bladelet technologies, new materials (bone, antler, ivory) and technologies (sawing, grooving, polishing, grinding, and perforating), tight standardization of forms, manufacture of personal ornaments (beads, pendants), transportation of shells and other materials over long distances, manufacture of musical instruments, drawings of animals, carved figures, and burials with grave goods. This period was also characterized by specialization and occupation in various regional habitats, accompanied by regional and historical change in cultures (Mellars and Stringer 1989; Mellars 1998).

Neanderthals and anatomically modern *H. sapiens* coexisted in the Middle East and Europe for more than ten thousand years before the Neanderthals died out, sometime between thirty-four thousand and twenty-seven thousand years ago. Lewis-Williams (2002) argues that Neanderthals and anatomically modern *H. sapiens* not only coexisted, but also interacted and alternately occupied the same rock shelters in the Dordogne in France. Furthermore, he argues that the Chatelperronian culture of some Neanderthals developed through selective imitation of certain features of the Aurignacian culture: specifically, borrowing of stone tool technologies and body painting, but not of representational art and elaborate burials. The symbolic representations in art and burials displayed by anatomically modern *H. sapiens* are universally accepted as evidence of modern linguistic capacity. Their absence in Neanderthals is consistent with the general view that they lacked fully modern language.

LANGUAGE EVOLUTION SCENARIOS

Most anthropologists believe that language evolved through at least two stages. Several anthropologists have proposed scenarios for proximate mechanisms of the earliest protolanguage. One of the major issues they have addressed is whether protolanguage was gestural or vocal, or both. All of these scenarios are continuity models focusing on an early stage of language evolution using ape models for the last common ancestor.

In one of the early attempts, Hockett and Ascher (1964) presented a model for deriving a productive "open call system," which they consider prerequisite for language, from a "closed call system" typical of nonhuman primates. They propose that open calls were achieved through blending of old vocalizations. Their protohuman ancestors are described as descendants of a gibbon-like ancestor, which during the Miocene was being pushed out of forests and was taking up hunting and food sharing.

Gordon Hewes (1973) was the first anthropologist to turn to ape language studies for clues to language evolution. He suggests that early hominids (known as hominins in modern taxonomy) communicated referential information by imitating animal behaviors, tool use, bodily functions of eating and drinking, etc. While noting that gestural aspects of language persisted, he acknowledges that spoken language has advantages over gestural language. He suggests that early vocalizations were also imitative and advocates the "mouth-gesture hypothesis," according to which mini-gestural imitation of high-low, front-back, etc., is made by the tongue.

Also in reference to ape language studies, Armstrong, Stokoe, and Wilcox (1994, 1995) propose a gestural model for the origins of syntax (grammar). These authors are linguists specializing in sign language. They argue that syntax arose from the breakdown (analysis) of the gestalt nature of gestures, which can be interpreted either as words or sentences, rather than from the combination (synthesis) of spoken words. The gestalt of an iconic gesture

Robin Dunbar's Gossip Scenario for Language Evolution

Robin Dunbar (1996) proposes that language arose in modern *H. sapiens* as a means for maintaining within-group social cohesion. Moreover, he argues that, functionally, gossip replaces grooming, which he believes is the major means for social cohesion among nonhuman primates. In contrast, he says that "the conventional view is that language evolved to allow males to do things like coordinate hunts more effectively" (79).

Dunbar says that the frequency of grooming is proportional to the size of the primate group. He also asserts that grooming time required to maintain bonds in a group larger than fifty to fifty-five individuals, which already entails spending 40 percent of the time in grooming, would be prohibitive. He suggests that some means other than grooming became important in hominids (now hominins) about two hundred and fifty thousand years ago as group size exceeded 150 individuals: "Could it be that language evolved as a kind of vocal grooming to allow us to bond larger groups than was possible using the conventional primate mechanism of physical grooming?" (Dunbar 1996, 78). Language has two key features favoring social cohesion: (1) speakers can talk to several people simultaneously, and (2) they can spread information among a large network of people. He also notes that whereas bonding may attract cheaters, language provides a means for assessing the reliability of potential allies. Dunbar cites with approval a colleague's idea that gossip allows people to track their own and others' reputations.

In a section called "testing the hypothesis," Dunbar says language should display the following design features appropriate to social bonding: (1) it should occur in groups larger than primate grooming clusters, and (2) it should be primarily devoted to the exchange of social information. His own research confirms both these expectations. Conversation groups commonly comprise four or five people (twice the size of primate grooming clusters), and people spend 60 percent to 70 percent of their conversation time on social topics.

Toward the end of his book, Dunbar introduces sexual selection theory and agrees that males use language to advertise themselves and compete for females. He notes that his own research reveals that although males and females both engage in social talk most of the time, male social talk is primarily about themselves, whereas female social talk is about other people. He also mentions the idea that language evolved to formalize ritual.

Given the group size aspect of his model, Dunbar speculates on what drove the increase in hominid group size. He notes that defense of food sources and protection from predation are commonly viewed as advantages of larger groups. He also suggests that in this case, predation may have been from raids by competing human groups. Finally, he suggests that nomadism may have favored alliances among groups.

Dunbar says that group size correlates with a measure of brain size called neocortical ratio (of neocortex to total brain size). He believes that group size is a key to social complexity, and that social coalition size is an even better measure of social complexity. Moreover, he says coalitions are maintained by grooming. He speculates that the enlarged brains of monkeys and apes were driven by selection for social intelligence. He describes his study of mean grooming group size in relation to total group size and neocortical ratio in twenty-two primate species: "Just as we expected, the mean size of grooming cliques correlates with both other measures" (1996, 68).

In regard to language origins, Dunbar dismisses the gestural models of language origins, arguing that "we can see precursors for almost all the features of human verbal communication in the vocalizations of Old World monkeys and apes" (1996, 140). He argues that the alarm calls of vervet monkeys, which refer to specific predators, constitute an "archetypal protolanguage" (141). Consistent with this view, he proposes the following three-stage model for language evolution: First, as group sizes increased in great apes, vocal grooming began to supplement physical grooming; second, as group size increased further, vocalizations began to take on social meanings; third, much later, symbolic content began to appear. He notes that after the origins of modern speech and the migration of modern *H. sapiens* out of Africa, languages began to diverge. In trying to explain this phenomenon, Dunbar suggests that group-specific dialects provide a cheat-resistant means for testing claims to group membership, and hence, for group support: "Someone who speaks in the same way you do, using similar words with the same accent, almost certainly grew up near you, and at least in the context of pre-industrial society, is likely to be a relative" (1996, 168).

Falk's Developmental Scenario for the Origin of Language

Dean Falk (2004) presents a developmental model for the evolution of protolanguage based on an analysis of mother-infant tactile, gestural, and vocal interactions in humans as compared to chimpanzees and bonobos. She locates the selection pressure for new hominin interactions in the loss of clinging ability in infants of larger-brained bipedal mothers during the transition from *Australopithecus* to early *Homo*. She argues that these changes stimulated hominin mothers to adopt new communication patterns as they periodically put their infants down while they foraged.

Specifically, she proposes that protolanguage had prosodic elements similar to those of baby talk or motherese, that is, musical speech directed toward infants by human mothers around the world today. She models the evolution of prelinguistic behaviors on the processes by which infants acquire the earliest manifestations of language in the context of motherese: "Since language acquisition today is universally scaffolded onto motherese, it is argued that selection for vocal language occurred after early hominin mothers began engaging in routine affective vocalizations toward their infants, a practice that characterizes modern women, but not relatively silent chimpanzee mothers" (Falk 2004, 3).

She notes that her "putting the baby down" hypothesis is a gradualist continuity approach to language evolution. She compares it to discontinuity approaches emphasizing the affective nature of nonhuman primate vocalizations as contrasted with human speech. These approaches postulate a sudden late appearance of language without a link with earlier communications systems of nonhuman primates.

Falk traces both similarities and differences between mother-infant communication in chimpanzees and bonobos versus humans. Noting that all three species display extended developmental periods compared to monkeys, she emphasizes that maternal support and carrying of infants up to two months of age is crucial for the survival of chimpanzee infants. She relates Plooij's report on the importance of infant whimpering in stimulating maternal support. She also relates work on infant-directed "hoo" vocalizations used by chimpanzee mothers to retrieve infants in contexts of travel and foraging. She relates that other infant-directed gestural and kinesic communications by chimpanzee mothers occurred in the contexts of carrying, cradling, nursing, eating, playing, and traveling. Likewise, she notes that mother-infant play bouts include turn taking in tickling and the play face and laughing. She reviews similar reports of bonobo mother-infant communications, noting that bonobos, unlike chimpanzees, engage in some vocal imitation. She concludes that "the nearly identical mother-infant tickling/laughter bouts of chimpanzees, bonobos, and humans provides some of the best evidence for the continuity hypothesis with respect to the evolution of mother infant communication" (Falk 2004, 17).

Turning to the literature on other-infant communication and early language acquisition in human infants, Falk focuses on the interplay between affective and symbolic elements. She documents the didactic role of melodic and exaggerated prosodic patterns in infant-directed vocalizations (similar to those of African apes) in engaging infants in turn taking, joint attention, intonation contours, and recognition of boundaries between words and sentences. She cites evidence that these interactions also help infants acquire meaning and grammar. She concludes that "the fascinating discovery . . . that infant-directed speech contains separate elements that serve to express emotions, on the one hand, and function as didactic devices, on the other, is consistent with the view that motherese evolved incrementally from largely affective ancestral vocal communications to its present highly complex form" (Falk 2004, 14). Reciprocally, infants display more intentional manual gestures, such as requests and offers of objects, in association with their first words.

Turning to hominin evolution, Falk hypothesizes that two differences between human and African ape development are related to hominin bipedalism and brain size and to protolanguage respectively: the slower rate of human locomotor development and the higher rate of vocalizations of human mothers. Reviewing evidence that increasingly difficult births were common by 1.6 million years ago as brain size reached 900 cc and the pelvis was fully bipedal, she argues that early *Homo* mothers were forced to park their motorically helpless infants as they foraged. At this point, she hypothesizes, mothers began to use "vocal rocking" to keep in touch and reassure their infants that "mommy is near," and over time words would have emerged from conventionalized melodies. Her model is based on evidence that differences in maternal care are significant factors in infant survival and that differences in vocal production and comprehension are hereditary. She joins other speculations on the nature of earliest words, approving the suggestion that "mama" may have been among them.

such as miming throwing or catching, for example, includes something acting, its action, and the result of the action on an object. When separated out, these elements correspond to an agent, an action, and a patient, in terms of the action grammar of some linguists. The iconic nature of signs, as contrasted with words, provides a clue to the difficult problem of deriving syntax and duality of patterning from protolanguage. Although declining to date their scenario, they note that *H. habilis* apparently had the brain anatomy necessary for this transition.

Merlin Donald (1991), a psychologist, has proposed a model for various stages of language evolution. His first stage, typical of apes and australopithecines, is characterized by episodic event representation, that is, memory of concrete episodes. His second stage, typical of *H. erectus*, is characterized by mimetic representations voluntarily enacted for social communication. His third stage, typical of anatomically modern *H. sapiens*, is characterized by narrative thought and mythic representations, that is, metaphorical representation of events using arbitrary symbols. His fourth stage, typical only of literate societies, is characterized by external visuographic representations (see chapter 10). Donald argues that arbitrary symbols arose out of conventionalized gestures typical of mimetic systems. He emphasizes the primacy of cognition in the evolution of language, noting that it also depended on the evolution of voluntary control of organs of speech production and comprehension.

In contrast to those focusing on proximate mechanisms of language production, several anthropologists have proposed scenarios for the ultimate, that is, adaptive, significance of language. The first two we discuss are developmental models drawing on characteristics of early language development in human children as compared to great apes. As such, they are also continuity models.

In line with their extractive foraging model of early hominid subsistence, Parker and Gibson (1979) argue that learning to identify, locate, and extract a wide variety of embedded foods required extensive apprenticeship in tool use and referential communication about the nature and location of these resources. Specifically, they propose that early hominids used imitation of food consumption and tool use and locomotion toward the source, along with such gestures as pointing, requesting, and object showing. These communicative patterns are typical of two-year-old human infants on the verge of speaking their first words. Subsequently Parker (1984, 1985) elaborated on this theme, proposing later stages of language evolution based on stages of child language development. Derek Bickerton (1990) also argued that stages of language evolution followed the same sequence as stages of language acquisition in modern humans.

In her developmental model, Dean Falk (2004) proposes that vocal language arose as an adaptation for distal communication between bipedal mothers and the infants. She proposes that protolanguage was similar to *motherese*, a simplified musical speech mothers around the world use to address their infants. Its turn taking and melodic structure provide a scaffold for language acquisition by infants today. Falk points to separate affective and symbolic elements in motherese as evidence of continuity between the emotional play of ape mothers and infants and the motherese of modern humans. She places the emergence of motherese sometime after 1.6 million years ago as hominins became fully bipedal and had more difficult births owing to their larger brains. Noting that differences in maternal care are significant in infant survival versus mortality, she suggests that mothers, forced to put down increasingly helpless infants as they foraged, used "vocal rocking" to soothe them.

In contrast to the preceding scenarios, the following scenarios focus on later stages of language evolution. Robin Dunbar (1993, 1996) proposes that language arose in modern *H. sapiens* as an adaptation for social cohesion. He suggests that gossip replaced grooming as a means for social cohesion as group size exceeded 150 individuals. Group size, he believes, is a good measure of social complexity and correlates with the ratio of the neocortex to total brain size. He proposes three stages of language evolution: first, as great ape group sizes increased, vocal grooming began to augment manual grooming; second, as hominid group size increased, vocalizations took on social meanings; and third, much later, symbolic content appeared. He also notes sex differences in speech consistent with the idea that men use language to show off.

Geoffrey Miller (2000) proposes a sexual selection model for the evolution of language. Like Darwin he views these as "arts of seduction," fitness indicators and means for assessing fitness (see chapter 7). Male competition, for example, can be seen in vocabulary size and in-group jargon. He notes that, consistent with handicap theory, courtship behaviors advertising fitness are notoriously wasteful (another scenario harking back to Darwin's idea about singing was proposed by Frank Livingstone [1972]).

Daniel Nettle (Morris 1967; Nettle 1999) in his group "marking hypothesis" notes that any attempt to explain the evolution of languages needs to

Noble and Davidson's Planning Scenario

In their book *Human Evolution, Language and Mind*, Noble and Davidson (1996) argue that the capacity for language and higher intelligence arose in *H. sapiens* as an adaptation for planning. They emphasize the saltatory nature of their model and their opposition to gradualist models including the notion of protolanguage.

After briefly reviewing studies of ape communication, in line with their saltatory approach, they dismiss apparent symbolic abilities of great apes as an artifact of human-ape interactions. After emphasizing their social construction approach to symbolism, they discuss the nature of codes, icons, and symbols and the relationship between perception and meaning.

Noble and Davidson argue against archaeological evidence others have used to bolster claims for the early, gradual evolution of language and culture. Then they focus on archaeological and anatomical evidence for the sudden emergence of language and culture in anatomically modern *H. sapiens* in the last sixty thousand years: "There are no signs of meaning for the producers of artifacts earlier than the first colonization of Australia about 60,000 years ago" (1996, 141). Specifically, they argue that Upper Paleolithic tools, art, shelters, burials, and colonization of islands are the first valid indicators of the kind of planning that requires language.

In their final chapter they argue that aimed throwing, pointing, gesturing, imitation, and joint attention were implicated in the emergence of language and culture. In summary, they say that "our argument overall is that appreciating the reality of signifiers has to arise in order that they can be made detachable from immediate contexts of association, carried off as it were (displaced) to other contexts, yet appreciated as continuing to refer to now absent signifieds. The trace, in 'freezing' the gesture [reference to J. J. Gibson] makes the signifier a new environmental entity so that its existence is created as something in itself, yet as existing in simultaneous relationship to both sign-maker and signified. Some such reorientation of attention away from the signified is needed in order to realize the existence of the sign as signifier. Howsoever that is accomplished, once done, the sign becomes usable symbolically, as a name" (Noble and Davidson 1996, 224).

---

explain why there are many languages rather than only one. To answer this question, he begins with a discussion of the evolution of cooperation, focusing on the problem of freeloaders. Turning to Durkheim, he argues that shared dialects act like shared rituals to create and mark shared identity within social groups, and indicate social boundaries between groups: "The problem of maintaining generalized reciprocity in a large group of mobile individuals is that of boundedness: a mechanism must be found to prevent insiders from defecting and outsiders from interloping. Creating a distinctive language would seem to be a way of doing this" (221).

Several anthropologists have proposed that language arose as an adaptation for planning (Gamble 1986; Whallon 1989; Parker and Milbrath 1993; Noble and Davidson 1996). Noble and Davidson argue for a discontinuity model of the sudden emergence of language and culture in anatomically modern *H. sapiens* about sixty thousand years ago. Appealing to a social constructivist model, they argue that Upper Paleolithic art and burial, as well as colonization of islands, are evidence for the kind of planning that requires language.

In a continuity model, Parker and Milbrath (1993) propose three stages in the evolution of planning based on the stages of development of planning in human children. Using this model to analyze archaeological trends, they also conclude that in the last stage, language-based declarative planning (typically developing in twelve- to sixteen-year-old children), evolved coincident with the evolution of hypothetical deductive thinking. They also argue that declarative planning was the basis for the Upper Paleolithic Revolution.

CONCLUSIONS ON LANGUAGE EVOLUTION

First, we should note that there is near unanimity among scenario builders that fully modern speech arose in the transition from archaic to anatomically modern humans between 160,000 and 60,000 years ago. There is less agreement, however, about the degree of change involved in this transition and its abruptness.

Second, most scenario builders who address earlier stages of language evolution specify or imply three stages of language evolution, at least one additional stage between the earliest and the last stages. These generally correspond to major stages of hominin evolution we have outlined.

Third, most scenario builders writing in or after the 1970s suggest some role for both gestural and vocal signaling, in line with great ape signing abilities and with the growing appreciation of the role that gestures play in spoken language. They differ, however, on the primacy of one modality or the other.

Fewer scenario builders have provided detailed accounts of the characteristics of early language. Armstrong, Stokoe, et al.'s (1994, 1995) scenario for the gestural origin of syntax is an exception. Falk's (2004) origin of words from emotive motherese is another. Likewise, Donald's (1991) scenario specifies increasingly complex forms of event representation, as well as stages of language evolution. All of these scenarios seem plausible, and are not necessarily in conflict.

Turning from the mechanics of language (proximate analysis) to its adaptive significance (ultimate analysis), we find a special challenge. Humans use language for many things, including explaining, teaching, persuading, warning, commanding, teasing, lying, reassuring, competing, and courting, as well as ritually transforming statuses. Therefore, distinguishing adaptive functions from side benefits (Williams 1966) is a major challenge. Moreover, the adaptive functions of language have probably changed during the course of evolution. We think the solution is to conceive of a sufficiently inclusive selection pressure at each stage.

We favor adaptive scenarios applicable to three stages of hominin evolution, discussed in reverse order of their evolution. First, we agree with the popular idea that fully modern language arose in anatomically modern *H. sapiens* as an adaptation for declarative planning. Adult-level language based on hypothetical reasoning allows people to construct and compare scenarios by imagining their outcomes without incurring the danger, wasted energy, and resources entailed in enacting them. Planning is a general mental phenomenon exercised in all-important subsistence, defensive, and reproductive domains, from hunting and warfare to courtship and parenting.

The planning scenario is consistent with the idea that language, like other forms of communication, is an adaptation for social manipulation rather than disinterested information transfer (Dawkins and Krebs 1978). According to this view, which we favor, language is a means for political persuasion and social control, as well as for seduction and courtship (Parker 1985; Alexander 1989; Miller 2000). As such it has been favored both by natural and sexual selection, for both male competition and display and as a basis for female choice, an idea proposed by Darwin. We also see creation and maintenance of social-linguistic boundaries as an important social function served by language divergence as described by Nettle (1999).

Second, we favor Dean Falk's (2004) motherese model for the origins of "vocal rocking" prelanguage (a different version of "singing") in middle-period hominins (*H. erectus*). As she notes, the increased brain size and full commitment to bipedal locomotion on open habitats would have resulted in greater dependence of nonclinging infants. Consequently, kin selection and parental manipulation would have favored enhanced distal communication between mother and infant in both auditory and visual channels. In addition to prelanguage, this probably led to selection for new facial displays, including everted pink lips and white sclera of the eyes to enhance communication (see chapter 7).

Third, in line with our preference for the extractive foraging subsistence model, we favor the early hominin scenario for the use of imitative gestures and movements to communicate to close kin the identity, location, and nature of a variety of embedded food sources (Parker and Gibson 1979). This is consistent with Hewes's (1973) emphasis on the primacy of gestural communication in early hominins. Both kin selection and parental manipulation would have favored this form of help to relatives.

The overall picture that emerges suggests continuities from one stage of hominin evolution to the next, beginning with great ape–like abilities in the last common ancestor, continuing through two or more transformations, and culminating in grammatical speech in anatomically modern humans. This time depth and multiple transformations of language evolution are consistent with the highly complex nature of the phenomena.

Studies of language production and comprehension reveal a wide range of cognitive skills entailed in language, including encoding and decoding spatial, temporal, and object properties, as well as classification, logical and mathematical skills, and social-emotional factors. This complexity of language is reflected in its widespread distribution in the brain (Elman, Bates, et al. 1996; A. Martin 1998). The close relationship between language and cognition can be seen, for example, in their parallel development. These factors all point away from models proposing the sudden, discontinuous appearance of an innate language module isolated from intelligence proposed by some evolutionary psychologists. They also point away from the idea of an instinctive language organ or module (Parker, Langer, et al. 2005).

For the sake of clarity, we discuss language, intelligence, and culture separately. Obviously, however, the three phenomena are intertwined and interdependent and have evolved in concert. After considering intelligence in the next chapter, we address cultures, and discuss relationship among the three phenomena in the final chapter.

9
Origins of Human Mentality

> There can be no doubt that the difference between the mind of the lowest man and that of the highest animal is immense. . . . Nevertheless, the difference in mind between man and the higher animals, great as it is, certainly is one of degree and not of kind. We have seen that the senses and intuitions, the various emotions and facilities, such as love, memory, attention, curiosity, imitation, reason, &c, of which man boasts, may be found in incipient, or even sometimes in a well-developed condition, in the lower animals
>
> —Darwin 1871, 494–95

In *The Descent of Man*, Darwin (1871) discusses the evolution of many aspects of human mentality, including tool use, curiosity, imitation, attention, memory, imagination, reason, sense of beauty, beliefs and superstitions, self-consciousness, and social and moral faculties. Reviewing both similarities and differences between human and nonhuman animal minds, Darwin argues for continuities in all elements and the gradual improvement of intellectual powers through natural selection.

Since Darwin, mental evolution has attracted fewer scenario builders than language evolution. Leaving broader aspects of this topic to psychologists, anthropologists have been more apt to speak of brain evolution and evolution of tool cultures rather than mental evolution. Everyone recognizes, of course, that humans are smarter than other primates, and that modern humans are smarter than human ancestors were, but practitioners of different specialties have focused on different aspects of mentality.

Paleoanthropologists have focused on brain size because bones are their stock in trade. Archaeologists generally have focused on tools and other elements of material culture since these are their stock in trade. Likewise, field primatologists have focused on tool use in wild chimpanzees since this behavior was discovered in our closest relatives (this focus was predictable given the anthropomorphic belief of earlier anthropologists that only humans made and used tools). In contrast, comparative psychologists have focused on social intelligence, particularly on imitation, self-awareness, and deception, for example, Machiavellian intelligence, among primates.

As anthropology has became more interdisciplinary, ideas about intelligence and especially studies of intellectual development have begun to influence primatologists, archaeologists, and paleoanthropologists. Beginning in the 1960s, practitioners of comparative psychology began applying a variety of human developmental models to the study of primate cognition: first linguistic models, then Piagetian and neo-Piagetian stages of physical, logical, and social reasoning, then stages of self-awareness, social roles, and theory of mind. All have been useful for comparing primate mentalities because they indicate greater and lesser degrees of intelligence, and yet describe species-specific patterns. These models have been practical because they focus on spontaneous motor behaviors rather than conditioned responses or verbal responses. They come close to meeting the ideal strengths of comparative frameworks, that is, comprehensiveness, flexibility, authenticity, discriminativity, accessibility, replicability, and compatibility (Parker 1990).

As we will see, some investigators have used these models to construct evolutionary scenarios. Before turning to these, and in line with our previous discussions, we begin with efforts to reconstruct the evolution of the cognitive abilities of the last common ancestor (LCA) of chimpanzees and humans.

In comparative terms, intelligence refers to the ability to solve problems, specifically, to invent and/or learn new ways to achieve goals, particularly to show the flexibility to apply new solutions in a variety of contexts (Kohler 1927). This suggests, for example, that species (e.g., Egyptian vultures and Galapagos woodpecker finches [Van Lawick-Goodall 1970]) that use only one kind of tool in one context for one task are less intelligent than species that use a variety of tools in several contexts for several tasks. If species engage in social learning of tool use through social imitation of novel behaviors, rather than simple application of existing behaviors, this also indicates greater intelligence. Social learning has the advantage of increasing the rate of transmission within and across generations (Russon and Galdikas 1995; Heyes and Galef 1996; Russon 1996). Likewise, we count as intelligent a species that displays the ability to deceive conspecifics in ways that involve taking into account their knowledge or lack of knowledge (Mitchell 1986; Byrne and Whiten 1992). If the species has a large brain and requires many years to become an efficient tool user, we would conclude that it is more intelligent than species with smaller brains and more rapid development. Finally, if a species shows related problem-solving abilities in several domains, we would be even more convinced of its greater intelligence.

INTELLIGENCE OF HUMAN ANCESTORS
What about the Last Common Ancestor's Intelligence?

Since the eighteenth century, in contrast to anthropologists, psychologists have recognized the unusual problem-solving abilities, especially the tool-using proclivities, of captive great apes (Kohler 1927; Yerkes and Yerkes 1929; Mitchell 1999). All the great apes use tools intelligently in captivity, but only chimpanzees (and humans) were known to use them in the wild. Recently, primatologists discovered that orangutans in Sumatra also use tools in the wild (Van Schaik, Fox, et al. 1996; Fox, Sitompul, et al. 1999; Van Schaik and Knott 2001). Since all of the great apes and neither the lesser apes nor the Old World monkeys use tools, it is likely that this ability arose in the common ancestor of all the great apes, who lived about fifteen million years ago (Parker and Gibson 1977; Van Schaik, Deaner, et al. 1999). If this is true, then the last common ancestor of chimpanzees and humans must have been a tool user.

Parker and Gibson's Extractive Foraging Scenario for Evolution of Cognition

Parker and Gibson (1979) propose a developmental scenario to explain the evolution of hominid (now hominin) subsistence, technology, and cognition. Based on a Piagetian model, they propose five stages of cognitive evolution associated with different feeding strategies:

1. prosimian stage, characterized by achievement of sensorimotor stages 1 and 2 based on simple object manipulation;

2. Old World monkey stage, characterized by achievement of sensorimotor stages 1 through 4 without secondary circular reaction—similar to that of nine-month-old human infants—based on occasional manual extractive foraging on encased and embedded foods;
3. great ape stage, characterized by achievement of sensorimotor stages 1 through stage 6 with elaboration of secondary circular reactions and deferred imitation and symbolic play similar to that of two- to three-year-old human children based on seasonal extractive foraging with tools;
4. a hominid stage characterized by achievement of sensorimotor stages 1 through 6 with elaboration of secondary and tertiary circular reactions and imitation, plus elaboration of preoperational notions of space and classification and seriation of objects similar to that of three- to five-year-old human children based on primary dependence on extractive foraging with tools;
5. an early *Homo* stage characterized by the additional achievement of a late preoperations stage understanding of straight lines and angles, and of one-to-one correspondences similar to that of five- to seven-year-old human children used in aimed throwing and simple stone tool production.

The authors propose that the later (fifth and sixth) stages of sensorimotor intelligence and early preoperational stages arose in great apes as adaptations for the seasonal feeding strategy of extractive foraging with tools on a variety of hidden and embedded foods. Discovering and extracting a wide variety of these foods (as opposed to a single food source) depends, among other skills, upon fifth and sixth sensorimotor stage understanding of the spatial relationship of enclosure and penetration as well as the cause and effect relationship of tool to the covering of the object, and mental representation of these relations.

They propose that early hominids elaborated and extended these abilities inherited from their common ancestor with great apes as they came to depend upon omnivorous extractive foraging as their primary subsistence mode, based on early preoperational abilities including pretend play, understanding of simple topological spatial relationships of penetration and enclosure, and simple functional grouping of objects.

Early hominids differed from Miocene great apes in both the scope and the frequency of their extractive foraging: "Hominid differentiation from the apes was based on a shift from secondary seasonal dependence (as in the case of chimpanzees) to primary year-round dependence of such tool-aided extractive foraging" (Parker and Gibson 1979, 371). They argue that

> the first hominids had a basic tool kit consisting of perishable wooden and other organic tools and unmodified stones, pounding stones for breaking open hard-shelled fruits and nuts, for cracking open scavenged bones for

marrow, for smashing open turtle shells, and so forth, digging sticks for excavating deep roots, tubers, and bulbs, and for digging for water, stabbing sticks for stabbing excavated fossorial animals, hitting sticks for knocking nuts, fruits, and seeds off bushes and trees; probes for termite fishing and ant dipping; leaves for cleaning and wiping grit from food, natural shell dippers for scooping up and drinking fluids from boles such as water, blood, and honey; and perhaps containers for collecting and transporting small extracted and gathered items such as grubs and nuts. (371)

Parker and Gibson argue that protolanguage and food sharing arose in the earliest as secondary adaptations to the extensive apprenticeship required for identifying and extracting a variety of embedded foods with tools. Food sharing represents an intensification of the ancestral great ape pattern of food sharing and apprenticeship still seen in chimpanzees. They propose that early hominids used simple referential imitation and gestural communication to family members about the location and identity of dispersed embedded foods. They note that modern human infants on the verge of speaking use similar gestures to point at and manually request and show objects (the gestural complex). They note that since modern great apes display the ability to use communicative gestures, this ability probably existed in their common ancestor with hominids.

Based on archaeological data, Parker and Gibson argue that *Homo habilis* elaborated on this pattern by adding new technologies of aimed throwing of missiles to stun or drive game into bogs or traps, and invention of worked stone tool technology for a major new form of extractive foraging—butchering and sharing large hunted and/or scavenged animals. They also argue that these hominids began to construct shelters. They argue that these new technologies were based on evolution of late preoperational abilities for straight-line construction, an understanding of angles and simple sectioning of solids, and construction of one-to-one correspondences characteristic of four- to six-year-old modern human children. They contrast the simple construction of nests, poor aimed throwing, and lack of tools used in hunting by chimpanzees with the ubiquity of construction and throwing games in young human boys.

After introducing a discussion of the evolution of development (heterochrony), Parker and Gibson conclude that "comparative data on primate development are consistent with the hypothesis that hominid intelligence evolved through a series of terminal additions of new abilities and a series of retrospective elaborations of abilities already present in rudimentary form in ancestral species" (1979, 380). Likewise, they relate stages of cognitive evolution to brain evolution and brain development in great apes, early and later hominids, and modern humans. They argue that there is some evidence that recapitulation occurs in brain development.

It is well known that great apes display all of the characteristics of intelligent tool users (Kohler 1927; Byrne and Whiten 1988; Parker and Gibson 1990; Parker, Mitchell, et al. 1994; Byrne 1995; Russon, Bard, et al. 1996; Byrne 1997; Parker, Miles, et al. 1999; Russon and Begun 2004). More significant for understanding the intelligence of the last common ancestor, it is well known that wild chimpanzees engage in a wide variety of tool uses in many contexts including feeding, grooming, self-protection, and play. Of all these contexts, feeding is by far the most common (McGrew 1992). Tool use is critical to fishing for termites and ants and for opening hard-shelled nuts, as well as procuring small tree-nesting animal prey, honey, brain juices, and water. Four common kinds of tools are hammer stones for cracking hard objects, wands or probes for extracting clinging insects, leaf sponges for sopping up water and juices, and leaves for cleaning dirt off of self. Nut cracking, for example, is a complex activity that entails considerable planning in finding suitable tools, transporting them to a suitable anvil site, and bringing the nuts there to be cracked. In the most complex cases, chimpanzees prepare anvils and even stabilize them with wedges (Goodall 1986; Boesch and Boesch-Acherman 2000; Matsuzawa 2001).

Moreover, these kinds of tool use require five to eight years of apprenticeship at the mother's side (Goodall 1986; Boesch and Boesch-Acherman 2000), and depend on social intelligence, both imitation of the mother's actions and teaching by demonstration by the mother (Boesch 1991b; Boesch 1993; Boesch, Marchesi, et al. 1994; Matsuzawa and Yamakoshi 1996). Finally, there is evidence that chimpanzees have locally variable cultural traditions of tool use (Whiten, Goodall, et al. 1999; see chapter 10).

In addition, chimpanzees show logical-mathematical knowledge as revealed in their ability to understand whole numbers and fractions and to add and subtract them (Boysen and Berntson 1990; Boysen and Capaldi 1993; Boysen, Berntson, et al. 1995). They have also shown the capacity to classify objects by two characteristics, for example, shape and size (Premack 1976), and to understand scale models of furniture and other objects in rooms (Kuhlmeier, Boysen, et al. 1999).

Last but not least, chimpanzees and other great apes have shown an ability to recognize themselves in mirrors (Gallup 1970; Miles 1994; Parker, Mitchell, et al. 1994; Patterson and Cohn 1994; Hart and Karmel 1996; Mitchell 1997), to imitate novel manual actions (Hayes and Hayes 1951; Russon and Galdikas

1993, 1995; Miles, Mitchell, et al. 1996; Russon 1996; Whiten and Custance 1996; Russon 1999), and to understand what their conspecifics know and do not know based on their visual access to information (Premack and Woodruff 1978; Woodruff and Premack 1979; Premack 1988).

A wide range of comparative studies agree that adult cognitive abilities of chimpanzees seem to correspond roughly to those of human children between the ages of two and four years (Premack 1976; Premack and Woodruff 1978; Russon, Bard, et al. 1996; Parker and McKinney 1999). Great apes differ from humans, however, in the unevenness in their levels of achievement and rates of development among the various cognitive domains, logical knowledge being more developed than physical knowledge (Langer 1996; Langer 2000a, 2000b), both of these being more developed than language (Parker and McKinney 1999.)

The data on abilities of great apes reveal both continuities and discontinuities between human and great ape mentalities. In Piagetian terms, they set a lower limit for the abilities of the earliest hominins at early preoperations (typical of two- to four-year-old human children). Preoperations involve symbolic, that is, representational thinking about simple causes such as tool-mediated opening of a nut and relationships such as similar shape, color, or size.

Conversely, data on modern humans set the upper limit for the average level of cognitive abilities of anatomically modern humans somewhere in the range of late concrete operations (typical of six- to twelve-year-old children) and early formal operations (typical of twelve- to sixteen-year-old children), depending on training. Concrete operations involve understanding of differing perspectives and reversible processes such as classifying the same group of objects by different characteristics. Formal operations involve generating and testing simple hypotheses such as the forces determining flight distance of an aimed missile (Inhelder and Piaget 1958; Dasen and Heron 1981; see table 9.1 for descriptions).

Anthropologists have tried to fill in this gap between humans and living great apes by inferring the cognitive abilities underlying subsistence modes and technologies revealed in archaeological and paleontological data on hominins but this provides limited and ambiguous information. As with subsistence modes, four kinds of comparative data on living great apes and foraging peoples have been used to infer cognitive abilities of earlier hominins: (1) stone tool technologies, (2) other material culture including shelters, fire,

Table 9.1. Summary of Piaget's Periods of Cognitive Development

Periods and Subperiods	Cognitive Domains		
	Physical Knowledge	Logical-Mathematical Knowledge	Social and Intrapersonal Knowledge
Sensorimotor period: birth to two years	Discovery of practical properties of objects, space, time, and causality	Sensorimotor, construction of logical relationships among objects	Discovery of interpersonal efficacy through circular reactions, and novel schemes through imitation
Preoperations period: two to six years; symbolic subperiod: two to four years; intuitive subperiod: four to six years	Discovery of immediate causes of actions and reactions	Construction of nonreversible classes	Construction of new routines and social roles through imitation and pretend play
Concrete operations period: six to twelve years	Discovery of simple mediated causes of actions and reactions	Construction of reversible hierarchical classes	Construction of more complex routines and roles based on rules
Formal operations period: twelve to eighteen years	Discovery of simple laws of physics through measurement and control of variables	Discovery of logical necessity through hypothetical deductive reasoning	Construction of universal rules and principles

notation, and artistic productions, (3) brain size and configuration from fossil endocasts, and (4) life history patterns and behavior inferred from dental and cranial characteristics of fossil skeletons. Before discussing these efforts, we briefly review the stages of hominin adaptation discussed in earlier chapters.

Intelligence of Basal Hominins

Since the earliest worked stone tools appear about 2.5 million years ago (Semaw, Renne, et al. 1997)—about four million years after the probable emergence of the earliest hominin—very little can be inferred about the mental abilities of basal hominins as compared to those of the LCA. Presumably, they showed some derived hominin characters, perhaps including bipedal aimed throwing of missiles (see chapter 5). Almost certainly, like the LCA, they used tools, but not worked stone tools (Mercader, Panger, et al. 2002; Panger, Brooks, et al. 2002). Given the brains and behavior of their descendants, it seems likely that basal hominins depended more on tool use than the LCA had. According to our favored model, they probably relied more on tool-aided extractive foraging than the LCA (see chapter 4). Certainly they were as intelligent, if not more intelligent, than the LCA.

Wynn's Scenario for Spatial Intelligence

Thomas Wynn (1989) proposes a model for the evolution of spatial competence in australopithecines and *Homo erectus* based on Piaget and Inhelder's stages of spatial understanding in human children. According to Wynn's analysis, early stone tools reveal three major stages of understanding in hominids (now hominins): The first stage, revealed in Oldowan tools, consisted of early preoperational thinking involving simple topological notions of space. The second, revealed in early Acheulean tools, consisted of late preoperational thinking involving simple notions of projective geometry. The third stage, revealed in late Acheulean tools, consisted of operational thinking involving simple Euclidean concepts. Preoperational thinking, typical of young children, involves some ability to perform actions mentally, that is, mental representation. Concrete operational thinking, typical of children older than six years, involves the ability to understand reversible events and conservation of simple properties such as quantity under transformations.

Wynn argues that an appropriate model for describing intelligence in prehistory must first define cognition in a manner suitable for comparing nonhuman primates, especially apes, with humans, and second, must assess end products of behavior, that is, stone tools. He believes that Piagetian theory fulfills these requirements. According to Wynn, early preoperational spatial understanding in human children focuses on intuitive qualitative topological relations including proximity, order, and continuity. Later preoperational spatial understanding focuses on perspectives, planes, and cross-sections. Operational spatial understanding focuses on Euclidean geometry of relative positions of objects in space, based on measurement, symmetry, and parallel axes. Although he finds this framework useful, he notes that some sequences in prehistory differ from those Piaget identified in development.

The sample of stone tools Wynn analyzed covers three time periods: The first was of choppers from Bed I and lower Bed II at Olduvai Gorge, dating from 2 to 1.6 million years ago. Regarding these, he concludes, "By modern standards, the spatial repertoire of two-million-year-old stone knappers was extremely limited. The hallmark of the Oldowan, the chopper required only a concept of individual blows or trimming scare (separation), a notion of positioning one next to another (pairs), and, perhaps, a concept of the edge (boundary) dividing faces of the tool. Scrapers . . . required the additional notion of sequences of blows arranged according to a constant direction (order)" (Wynn 1989, 59–60). They also reveal some attention to the shape of the edge.

The second sample was of cleavers from upper Bed II at Olduvai Gorge and West Natron, dating from about 1.5 to 1 million years ago. Regarding these, Wynn concludes that in addition to earlier concepts, the hominids of this period added the notion of interval or constant quantity of space: "1.2 million-year-old assemblages include very circular discoids and very round spheroids. These require at a minimum, a notion of diameter or radius, which are both constant amount of space used as an internal reference. . . . A second spatial notion that appears at this time is that of symmetry. It was not a full-scale euclidean symmetry of mirrored congruency, but it did include the mirroring of shapes" (1989, 60). He notes that these knappers no longer focused solely on edges, but also on shapes.

The third sample was of hand axes from Ismila, dating from about three hundred thousand years ago. Wynn describes an Ismila hand ax as demonstrating symmetry across the midline, the profile, and all the cross-sections, forming intersecting lines defining three-dimensional space. Their regular cross-sections require precorrection and understanding of interchangeable points of view. Regarding these tools, Wynn concludes, "By 300,000 years ago there is evidence for an essentially modern concept of space, one that extended beyond the individual framework of the artifact and organized space as a whole. . . . Perhaps the most critical new spatial concept is the understanding and coordination of multiple points of view. . . . A second concept to appear by 300,000 years ago is that of 'euclidean' space, that is, a space definable by a three-dimensional coordinate grid" (1989, 61). Tool production in this period is also characterized by an understanding of part-whole relations.

Wynn argues that the archaeological sequence resembled the ontogenetic sequence except that the development of a general frame of reference in space occurred without a stage of competence in producing parallels. It was also characterized by an early appearance of the notions of interval, shape, and mirroring. Both sequences involve giving up local, internal frames of reference for a general, external framework.

Overall, Wynn notes that great apes display preoperational thinking similar to that of Oldowan and even early Acheulean toolmakers. He also notes that whereas his analysis focuses on minimal competence in tool production, these hominids may have displayed higher intelligence in other archaeologically "invisible" realms such as social behavior. In contrast, he says, late Acheulean toolmakers three hundred thousand years ago already displayed operational thinking. This, he says, "places an essentially modern intelligence much earlier than the time of the first appearance of hominids with completely modern anatomy, that is, *Homo sapiens sapiens*" (1989, 89).

He explicitly lumps together concrete and formal operational intelligence, arguing that formal operational intelligence is "virtually invisible archaeologically." Moreover, he doubts that it is the end product of human evolution. Wynn says little about the selection pressures acting on intelligence, but favors the notion that they were social.

Intelligence of Early Human Ancestors

As just indicated, the first worked stone tools appear at about 2.5 million years ago (Semaw, Renne, et al. 1997). Chopper tools associated with cracked marrow bones of antelope were discovered with *A. garhi* (Asfaw, White, et al. 1999), indicating that extractive foraging had been extended to using tools for scavenging bone marrow and brains, as well as for extracting various other animal and vegetable foods, perhaps including underground storage organs (see chapter 4). Aimed throwing of missiles to ward off predators and competitors was also likely. This suggests that they had consolidated and extended preoperational intelligence at least in the physical domains of space and causality. Likewise, elaborated social intelligence in imitation and pretend play would have been favored by extended apprenticeship in new foraging skills.

Intelligence of Middle-Period Human Ancestors

Fully bipedal *H. erectus/ergaster* first appeared about 1.9 million years ago. As discussed in chapter 4, these creatures had larger brains (about 900 cc), produced more advanced tools (biface hand axes and other Acheulean tools), and ate larger prey animals than their ancestors. They were the first hominins to migrate out of Africa. Probably they were still primarily scavengers rather than hunters, and almost certainly excavated and ate underground storage organs (USOs). They constructed shelters and they may have used fire (see chapter 4). As discussed below, Piagetian analysis of Acheulean tools suggests that in their later stages, middle-period hominins had achieved concrete operational intelligence (Wynn 1989). Few other artifacts exist to guide us.

**Intelligence of Recent-Period Human Ancestors:
Archaic and Anatomically Modern Humans**

By about two hundred thousand years ago, large-brained archaic *H. sapiens* had spread from Africa throughout the Old World. These descendants of *H. erectus* were producing more advanced Middle Stone Age (MSA) tools in Africa and Middle Paleolithic tools (for example, Mousterian tools including prepared cores, scrapers, knives, and borers) in Europe. They were hunting larger game, cooking foods, and making clothing and body adornments (see chapter 4). Despite this, in Europe their technology remained static for more than one hundred thousand years. In Africa, in contrast, during this period there was a more gradual "nuanced" transition from MSA to Late Stone Age (LSA) culture (McBrearty and Brooks 2000; see chapter 4).

This transition occurred sometime between two hundred and fifty thousand and sixty thousand years ago, before the appearance and geographic radiations of anatomically modern *H. sapiens* out of Africa and around the world into a variety of new ecosystems. Diffusion of technological, artistic, and social innovations indicative of modern mentality accompanied this migration. New technologies included composite tools, pressure-flaked projectile points, spear throwers, bone and antler needles, ceramic vessels, and new modes of food preparation, including grinding, drying, and storage. New social adaptations were revealed in ritual burials (such as status differences in clothing and jewelry), in mobilary art in large seasonal aggregations of hunting populations, and in artistic recording of activities of humans and game animals in cave paintings (Mellars and Stringer 1989; McBrearty and Brooks 2000). By ten thousand years ago, modern humans had domesticated plants and animals and begun to organize cities (see chapter 4 for discussion).

It is important to note in this context that paleoecologists have discovered that the past fourteen million years (the Miocene and Pliocene epochs) have been marked by deteriorating climates. Even more striking in relation to human evolution, the climate of the past 1.6 million years (the Pleistocene epoch) has been marked by "rapid, chaotic, and ongoing reorganizations of ecological communities" (Richerson and Boyd 2000). Richerson and Boyd attribute the evolution of human intelligence to selection pressures generated by these climate challenges.

SCENARIOS FOR THE EVOLUTION OF INTELLIGENCE

Competing Scenarios for Primate Intelligence

There has been a long-running dispute among primatologists over the primacy of social versus technological intelligence, or more recently, their interdependence. A closely related dispute continues over the independence, specialization, or "modularity" versus the generality, integration, or interdependence of various elements of intelligence in both human and nonhuman primates. Both these disputes are reflected in various scenarios of hominin intellectual evolution.

Most primatologists trace the social intelligence hypothesis to Nicholas Humphrey's (1976) essay on the social function of intellect, preceding that to Allison Jolly's essay on lemur intelligence (Jolly 1966) and Michael Chance and the Meads' book (Chance, Mead, and Mead 1953). Likewise, Frans de Waal's (de Waal 1983) study of political jockeying for dominance among captive chimpanzee males clearly illustrates their capacity for deception and negotiation. In any case, Richard Byrne and Andrew Whiten have greatly enhanced this perspective with their works on deception and Machiavellian intelligence (Whiten and Byrne 1986; Byrne and Whiten 1988, 1990; Whiten and Byrne 1997). Adherents to this school of thought argue that human and nonhuman primate intelligence arose as an adaptation for social manipulation.

In contrast, others have argued that human and nonhuman primate intelligence arose as an adaptation for feeding (Milton 1988) or feeding technology. Parker and Gibson (1977) argued that intelligent tool use arose as an adaptation for extractive foraging with tools on a variety of embedded and encased food sources including ants and termites, nuts, and honey. Specifically, they proposed that this form of extractive foraging arose in the common ancestor of great apes (and independently in cebus monkeys) as an adaptation for foraging on high-energy embedded fallback foods during dry seasons (Parker and Gibson 1977).

Later Parker (1996) elaborated this scenario to include evolution of true imitation, self-awareness, and demonstration teaching in the common ancestor of great apes. According to this elaboration, these great ape capacities, typical of early preoperational knowledge of two- to four-year-old children, arose as secondary adaptations for apprenticeship in intelligent tool use in extractive foraging. In other words, social and technological intelligence coevolved in great apes. See table 9.2 on hypothesized intellectual functions in chimpanzees.

All of these scenarios are compatible with the observation that brain size in many mammals increased through the past sixty-five million years (in the Cenozoic), and especially in the past 2.5 million years. This increase was especially noticeable among social mammals. This trend corresponds to increasing deterioration and complexity of climate (Richerson and Boyd 2000).

Scenarios for Hominin Intelligence

Anthropologists disagree about when humans achieved their modern intellectual abilities. Some, like Richard Klein (1999), argue that modern human mentality evolved about sixty thousand years ago and allowed the LSA achievements. Others, like McBrearty and Brooks (2000), argue that modern intelligence arose coincident with *H. sapiens* about two hundred and fifty thousand years ago in the African Middle Stone Age. They believe that the Late Stone Age achievements resulted from long-term trends in population growth and crowding led to intensification. Thomas Wynn (1989) argues that humans achieved the capacity for modern intelligence in the Late Stone Age, about 1.4 million years ago.

Table 9.2. Hypothesized Use of Various Cognitive Domains in Daily Activities of Chimpanzees

Activity	Logical-Mathematical	Physical	Social	Symbolic
Preparing implements and foods for use	Classification of objects; reckoning quantities	Physical causality and spatial relations in tool making and tool use		
Hunting game		Physical causality and spatial relations in predicting escape routes	Awareness of positions and activities of self relative to others	
Political activities	Keeping track of quantities and categories of goods and services rendered	Use of objects in intimidation displays and in giving gifts to others	Awareness of activities and appearance of self to others	Possible use of symbolic gestures in greeting
Coordinating progressions		Spatial maps of key resources and of group members; ability to take detours	Awareness of effect of activities of self on others	Possible marking of directions of movement
Nest building		Use of causal and spatial relations to construct adequate sleeping platform		
Teaching offspring		Use of causal and spatial relations in tool use	Awareness of effect of own actions on others	
Courtship and mating			Awareness of appearance in directed sexual displays	Possible iconic gesturing of preferred position

Source: Parker, Sue Taylor, and Michael L. McKinney. 1999. *Origins of intelligence: The evolution of cognitive development in monkeys, apes, and humans*, p. 213. Copyright © 1999 The Johns Hopkins University Press. Reprinted with permission of The Johns Hopkins University Press.

Mithen's Architectural Scenario for Mental Evolution

Steven Mithen's (1996) scenario of mental evolution covers "four acts," beginning six million years ago with the ancestral ape and ending ten thousand years ago with the agricultural revolution. After arguing for "cognitive archaeology," Mithen addresses the "architecture of the mind" by reviewing various psychological models. Acknowledging his inspiration by Thomas Wynn, he touches on Jerry Fodor's idea of mental modules, Howard Gardner's multiple intelligences, and the "Swiss Army knife" model of innate mental modules. He settles on Annette Karmiloff-Smith's developmental model of domain-specific mentalities—language, intuitive psychology, intuitive biology, and intuitive physics—all of which develop from generalized intelligence in infancy into domain-specific mentality in adults.

Based on his metaphor of the mind as a cathedral, Mithen proposes three phases of mental evolution recapitulated during modern human development: (1) a central nave dominated by generalized intelligence based on general-purpose (trial-and-error and associative) learning and decision-making rules; (2) addition of isolated chapels housing specific domains of social intelligence, natural history intelligence, and technical intelligence; and (3) interconnection of chapels conferring cognitive fluidity based on flow of ideas between domains.

Viewing chimpanzee behavior as the best model for ancestral apes and the earliest human ancestor, Mithen assesses their intelligence as indicative of act 1. He concludes that chimpanzee tool use, mental maps, group hunting, and symbol learning can be explained by general intelligence rather than specialized modules in language, technical intelligence, or natural history intelligence. In contrast, he concludes that their social behavior reflects a specialized social intelligence. At this point and again at the end of his book, Mithen speculates on the early origins of a social intelligence module. Based on work on vervet monkeys, he argues that it arose in the common ancestor of Old World monkeys by about thirty-five million years ago.

Then for act 2, he turns to the mind of the first stone tool makers, early *Homo* and possibly *Australopithecus*, 2.5 million years ago. Contrary to Wynn, he argues that Oldowan toolmakers had greater technological abilities than chimpanzees. Indeed, he argues that they showed the first evidence of specialized technological intelligence. Mithen also argues that they displayed ambiguous evidence of natural history intelligence for predicting resource locations. Nevertheless, he believes that general intelligence continued to play an important role in both tool making and foraging. Based on their increased brain size and configuration, however, he infers that their social intelligence had become more complex and powerful.

Then for act 3, scene 1, he turns to *Homo erectus* (1.8 million years ago), arguing that they display the first evidence of distinct domains of technology and natural history. He bases this judgment first on their production of Acheulean hand axes, characterized by symmetry and imposed form; second, on their ability to colonize new regions of the Old World; and third, on their enlarged brain size.

For act 4, scene 1, about two hundred thousand years ago, Mithen turns to Neanderthal. He puzzles over the apparent paradox of the modern brain size of Neanderthal combined with their limited technology, particularly the absence of bone, antler, and ivory tools, special-purpose tools, and multipurpose tools. Likewise, he puzzles over the lack of technical variation across space and time despite colonizing new habitats. In contrast, he argues that their large brain size indicates high social intelligence, but he puzzles over their small group size, and their lack of ritual and personal decoration.

Mithen's solution to these puzzles is to argue that the barrier between technological, natural history, and social intelligences was circumvented by general-purpose intelligence. He also argues based on brain features and hyoid bone configuration that Neanderthal had language, but that it was limited to the social domain. Finally, he argues that these creatures lacked introspection.

For act 4, scene 2, Mithen discusses the "big bang" of Upper Paleolithic culture that occurs suddenly about sixty thousand years ago with the appearance of art, ritual burials, personal ornaments, and complex, regionally and temporally evolving tool cultures. He says these new features result from cognitive fluidity achieved by integration of social and natural history domains, followed by integration of technological intelligence. In his architectural metaphor, doors between the chapels were opened.

Mithen focuses very little on selection pressures. He does argue that the greatest selective pressure on brain enlargement was social competition, and the greatest pressure on cognitive fluidity was provisioning females with food, attendant upon brain enlargement and increased infant dependency.

In this chapter we focus on five scenarios aimed at explaining some or all aspects of human mentality. In the tenth and final chapter, we focus on a scenario aimed at explaining the evolution of mentality in relation to culture.

First, we focus on two technological scenarios, both of which use developmental stages described by the child psychologist Jean Piaget. Parker and Gibson (1979) extended their extractive foraging model to early hominids (known as hominins in modern taxonomy), arguing that australopithecines began to depend more heavily on tools to extract a wider variety of embedded foods as their habitat dried. They also argued that these creatures had extended apprenticeships in tool use and communicated referentially through imitation (see chapter 8). They argued that these activities extended and elaborated the early preoperational intelligence of these creatures.

They further proposed that *Homo habilis* elaborated on these activities, engaging in more complex activities including production of pebble tools, butchery, food sharing, shelter construction, and aimed throwing of missiles. They argued that these entailed intuitive, that is, late preoperational, intelligence characteristic of four- to six-year-old children (Piaget and Inhelder 1967, 1969). Specifically, that chopper stone tool production involved an understanding of sharpness, angle, and transmission of force through object contact, that butchery and food sharing involves one-to-one correspondence between food and recipient, that aimed throwing requires constructing a straight line between thrower and target.

Finally, they argued that from apes to humans intelligence evolved through a series of terminal additions of higher stages of cognitive development (as well as retrospective elaborations). Therefore, they argue, mental development in modern humans recapitulates the stages of evolution of intelligence.

Likewise, the archaeologist Thomas Wynn has used Piagetian concepts to analyze Oldowan and Acheulean tools in light of the spatial concepts entailed in their production (Wynn 1979, 1989; Wynn and McGrew 1989). Specifically, he argued that chopper tools required only typological concepts of separation of individual blows, order of individual blows, and edge or boundary. He concluded that Oldowan tool production in early hominids (known as hominins in modern taxonomy) depended upon preoperational spatial concepts typical of preoperational intelligence of preschool age children (Piaget and Inhelder 1967). He also argued that these concepts are within the grasp of great apes.

In contrast, Wynn argued that Acheulean tool production, at least in its later phases, depended upon projective and Euclidean concepts, particularly the idea of perspective that emerges in concrete operations. Specifically, he argued that by three hundred thousand years ago, hand axes demonstrated an understanding of symmetry across the midline, profile, and all cross-sections, indicating perspective and interchangeable points of view. Finally, he argued that once hominids had achieved operational-level intelligence, all their subsequent technological achievements were strictly cultural. In contrast to Parker and Gibson, Wynn does not believe that formal operations depend upon specifically evolved abilities.

The scenario of Steven Mithen (1996), also an archaeologist, traces the evolution of the human mind through a series of transformations beginning with hardwired specialized modules fifty-six million years ago, followed by general intelligence thirty-five million years ago with the origin of Old World monkeys, then back again to specialized intellectual domains in a series of human ancestors culminating in a new generalized cognition with fluidity in modern *H. sapiens*. Using Karmiloff-Smith's model of human development from generalized intelligence in infancy to domain-specific in adulthood, Mithen argues that this developmental pattern recapitulates the evolutionary sequence.

Specifically, he argues that six million years ago, human/ape ancestors, like chimpanzees, used general-purpose intelligence to solve technological and ecological problems, but used a specialized social intelligence module for social problems. Beginning 2.5 million years ago with early *Homo*, Oldowan stone toolmakers displayed some evidence of specialized technological intelligence, but continued to rely on general intelligence. Based on their increased brain size and in accord with Dunbar's rule, Mithen argues that these creatures had more powerful specialized social intelligence than their ancestors. Based on their Acheulean tools and migration out of Africa, he argues that *H. erectus* at 1.4 million years ago was the first to display distinct technological and natural history intelligences. In reference to the Upper Paleolithic, he argues that modern *H. sapiens* sixty thousand years ago were the first to show true mental flexibility based on communication among specialized intelligences.

In terms of his cathedral metaphor, the first phase of generalized intelligence in human ancestors was like a central nave, specialized modules were like isolated chapels built on the side, and cognitive fluidity was like interconnection of these chapels. Mithen's oscillating model is his solution to the apparent paradox of the relatively large brains of *H. erectus* and later Neanderthal coexisting with tool cultures that remain static for one million and one hundred thousand years, respectively.

In contrast to these scenarios, two more recent scenarios emphasize intelligence as an adaptation for social manipulation. Geoffrey Miller's (2000) scenario focuses on the good side of human mentality. Richard Wrangham and Peterson's (1996) scenario focuses on the other, dark side of human behavior. Unlike the technological scenarios, neither of these social scenarios focuses on the proximate mechanisms of hominin intelligence. Likewise, neither of these is a developmental model.

Wrangham and Peterson's "Demonic Male" Scenario for the Evolution of Violence and Intelligence

Wrangham and Peterson (1996, 47) propose that "our own male-bonded, war-shaped societies have evolved for whatever reasons they did in chimpanzees during the time of our common ancestor or earlier. Five million years ago, this suggests, there were killer apes." They devote the remainder of the book to showing how violent behaviors benefit their perpetrators. Their theoretical framework is sexual selection; their methodology entails tracing continuities and step-by-step transitions between chimpanzee and human ancestors.

First, they turn to the twin questions of where the ape ancestors of humans lived and what they ate. Their answer is that our earliest ancestors lived in woodland habitats and subsisted on roots during the crunch season when fruits, leaves, nuts, and seeds are unavailable: They note the high density of these foods in woodland habitats, and the frequent association between fossil hominids and fossil mole rats, which depend entirely on this food source. Wrangham and Peterson also note that modern foragers, as well as one woodland-foraging group of chimpanzees, use these foods. The authors also note that these chimpanzees carry their excavated roots considerable distances bipedally before consuming them in protected settings.

Second, Wrangham and Peterson turn to the practice of chimpanzee raiding. They say that there is a common theme of killing members of neighboring groups of their own species, in other words, that killing and warfare are not unique to our species as previously thought. The rest of the section on raiding is devoted to comparing patterns of chimpanzee raiding and warfare in primitive human societies, focusing in detail on the horticultural Yanomamo of Venezuela and Brazil. In both species, the raids are stealthy, and killing and taking fertile females is the goal. Moreover, successful warriors have greater reproductive success than other men.

The authors note that surveys of primitive warfare suggest that as among the Yanomamo, lethal raiding is the commonest form of warfare in other societies. Moreover, surprise is the single most common principle. They cite a study of nonviolence in relatively egalitarian hunters and gatherers, showing that 64 percent engaged in warfare every two years, and 26 percent less often.

Third, Wrangham and Peterson examine the argument that warfare is culturally determined by examining popular images of paradise. They note that Paul Gauguin's Marquesans lived in a constant state of suspicion, hostility, and warfare, that Herman Melville's Typee are characterized by intermittent warfare and cannibalism, and that Margaret Mead's Samoans had persistent, sporadic warfare culminating in tearing out the hearts of captives and burning others. In addition, Samoans also showed a virginity cult, rape, and extreme jealousy. The authors note that the paradisiacal images of the South Seas created by these three had the common theme of focusing exclusively on females; they conclude that there is no paradise and never has been.

Fourth, Wrangham and Peterson note that their argument about the common ancestry of chimpanzee and human violence is an argument about the violent temperament of males. Therefore it also contradicts the popular idea that gender is culturally determined. In questioning this assertion, the authors cite three lines of historical and ethnographic evidence. First, they cite the rarity of female roles in warfare; second, the vastly greater frequency of violent crime by males as compared to females; third, the universally greater political power of males. Along the way they debunk claims of female warriors and of early matriarchies. They conclude that patriarchy is "worldwide and history wide" because it serves the reproductive interests of males.

Fifth, the authors survey studies of relationship violence in male humans and other great apes. Chimpanzee males fight with other intragroup males, competing for dominance rank and for access to fertile females. They also batter and rape females in their attempt to monopolize them sexually during their most fertile periods. Rape is common among male orangutans that have not achieved full adult male body size and vocal and bodily displays that allow them to hold territories and repulse other males. Male combat and infanticide are common among male gorillas during attempts to drive out harem-holding males or to steal females from them. The authors emphasize that all these forms of aggression are strategies of male competition and/or female control that increase male reproductive success.

At this point in their argument, Wrangham and Peterson suggest that both female vulnerability and high intelligence of males favor demonic behavior. Specifically, they argue that intelligence multiplies the strategies and tactics of social/sexual manipulation: "Intelligence turns affection into love and aggression into punishment and control. Far from being the mindless expression of some deep and bizarre ancestral trait then, the intense violence of apes arises partly from the very elaboration of these cognitive abilities" (1996, 152).

The authors then turn to the question of why great apes evolved these patterns whereas most other species did not. Surveying the few species that display lethal raiding (hyenas, lions, and wolves), they note that infanticide is the most widespread form of intraspecies killing because it is less risky for the perpetrators. They argue that harem social structures evolved in many species to protect infants from this common threat. Introducing the notion of cost versus benefit, they argue that killing will evolve whenever it is cheap enough and the benefit great enough.

They introduce the term "party-gang species" to describe species like chimpanzees and humans and hyenas with coalitionary bonds and variable party size. They argue that party-gangs cannot afford to live in stable groups year round because their food sources vary widely in distribution and scarcity throughout the year. When food is scarce and widely distributed they disperse into small groups, when it is abundant and concentrated, they congregate in large parties. "If you're a party-gang species living in rivalry with neighbors, a chance to kill safely tends to pay off for the same underlying reason. It weakens the neighbors" (p. 167). Therefore, they say, these species are demonic. They argue that male bonding is favored in these species because as party sizes decrease individuals become more vulnerable to attack. Males travel farther and have more time to interact because the cost of extra travel is less for them than for females.

Wrangham and Peterson end their argument with reflections on the contrasting social behavior and temperament of bonobos, the closest living relatives of chimpanzees. Despite many similarities with chimpanzees, bonobos differ significantly along the demonic dimension. Females are co-dominant with males, the highest-ranking male and female are equal. Mothers and sons have a close bond and a mother's support is critical for attaining dominance. Females bond and cooperate against males whereas males do not. Sex is the basis for female bonding. When females change groups at adolescence, a resident female who initiates sex with them first accepts them. Sexual competition among males is low key. Intergroup relations range from tense to placid and even friendly.

The authors ask how bonobos diverged so sharply from their close relatives the chimpanzees and humans: The first key is female cooperation. The other key, they say, is lack of intercommunity violence. These traits arose in the context of the larger, more stable size of bonobo parties. This is possible because their diets, a combination of chimpanzee and gorilla diets, include more stable, abundant foods. Stable party size allowed time for supportive relations among females to develop (which is impossible for chimpanzee females who spend most of their time alone with their offspring). Stable party size also prevents imbalances of power when groups meet, and hence lethal raiding.

They end by pointing out the double bind female humans and chimpanzees face when they oppose male violence: "If they support each other too much, they become vulnerable to losing what they want, the investment and protection of the most desirable men. There is competition between women for the best men, and it can break the unwritten compact among women" (p. 243). Sadly, they conclude, "The problem in both human and ape history is that political power is built on physical power—and physical power is ultimately the power of violence, or its threat" (p. 243).

Geoffrey Miller's Sexual Selection Model for Hominid* Brain Evolution

Geoffrey Miller (2000) uses sexual selection theory to explain the evolution of human mental abilities, including the capacity for morality, music, art, culture, and language, as well as intelligence. He begins with a review of Darwin's theory of sexual selection, according to which male competition and female choice explain sex differences that cannot be explained by natural selection. Like Darwin, he notes that sexual selection also explains many features that are unique to particular species and populations.

Miller traces controversies over sexual selection theory and its subsequent fall and resurrection. He also discusses such elaborations of sexual selection as Fisher's runaway sexual selection and fitness indicators and Zahavi's handicap principle. In Miller's view, the handicap principle, according to which costly ornaments are reliable fitness indicators, helped the revival of sexual selection theory (2000, 65) in the 1990s.

---

*Hominids are now known as hominins.

After noting that the social sciences have rejected evolutionary explanations for human creativity, he suggests that their limited success in explaining human behavior may spring "from over looking Darwin's crucial insight about the importance of sexual competition, courtship, and mate choice in human affairs" (Miller 2000, 67). He argues that "the intellectual and technological achievements of our species in the last few thousand years depend on mental capacities and motivations originally shaped by sexual selection" (21). He says that unique human abilities cannot be explained by natural selection because they do not show obvious survival value: "Nobody has been able to suggest any plausible survival payoffs for most of the things the human mind is uniquely good at, such as humor, story-telling, gossip, art, music, self-consciousness, ornate language, imaginative ideologies, religion, and morality" (18).

After exploring runaway selection, Miller concludes that it does not explain human mental evolution. The reason it fails is that brain enlargement has been a repeated, long-term trend while runaway selection is arbitrary and unlikely to produce consistent change in one direction. Moreover, it probably would have produced greater sex differences in brain size and mentality than humans display. Therefore, Miller argues that mutual choice rather than runaway selection has driven sexual selection in humans.

He notes that whereas sexual selection generally changes male courtship and competition characters as a consequence of polygynous mating by few males, even monogamous mating can produce sexual ornaments through "fitness matching" because less fit couples have lower reproductive success than fitter couples. Fitness matching produces greater fitness spread and hence diversity than polygynous mating. It also produces less sexual dimorphism. Likewise, condition-dependent courtship behavior produces flexible displays such as those typical of humans.

Miller's model of mental evolution through sexual selection focuses on *fitness indicators* (biological traits that evolved specifically to advertise an animal's fitness [2000, 103]): "This theory of fitness indicators suggests that much of human courtship consists of advertising our physical fitness and mental fitness to sexual prospects" (111). He notes that, consistent with the handicap principle, this advertising entails "prodigious waste." Costs include time, energy, and susceptibility to predators. Indeed, Miller argues that great wastefulness, variability, and heritability characteristic of human courtship behavior are exactly the features we should expect of fitness indicators.

Miller notes that human males are almost unique among primates in combining courtship and parenting; therefore, he argues, female choice must have combined with children's choice to reward parental behavior even toward nonkin (2000, 203). Children's choice made males better fathers. Unlike some evolutionary psychologists, Miller believes that hunting and food sharing with females and their offspring—as well as defense of resources—was driven by showing off fitness. He argues that this is consonant with the behavior of living hunter-gatherers.

Likewise, he discusses how such human bodily features as enlarged primary and secondary sexual characteristics appearing at puberty evolved through sexual selection as fitness indicators. Bodily symmetry seems to be preferred as an index of developmental stability. He sees the evolution of prolonged foreplay and copulation, and concealed estrus and orgasm in human females, as driven by female choice. Turning to modern human behavior, Miller says, "The signal difference between modern life and Pleistocene life is that we have the social institutions and technologies to benefit from the courtship efforts of distant strangers" (2000, 430).

Miller's main focus, however, is on intelligence, imagination, art, and language as "arts of seduction" favored as fitness indicators. He notes that innate human bodily ornamentation is elaborated by cosmetics, hairdos, clothing, and jewelry. On a larger scale, we humans ornament our residences, our tools, our means of transport, our cattle, and whatever other objects and spaces we command with art, music, and any other means to display our status and our fitness.

The most original aspect of Miller's model is his attribution of the human capacity for language, sympathy, morality, art, and science to sexual selection. Regarding morality and charity, which cannot be attributed to nepotism or reciprocal altruism, he says they often look like just another form of wasteful, showy display. Likewise, regarding sympathy he says "much of human courtship consists of sympathy displays. We show kindness to children, we listen to sexual prospects enumerating their past sufferings. The development of emotional intimacy could be viewed as the mutual display of capacities for high levels of sympathy" (2000, 330).

Regarding language, Miller says that in addition to such other functions as social coordination and teaching, "language puts minds on public display where sexual choice could see them clearly for the first time in evolutionary history" (2000, 357). Moreover, in addition to the status display functions of language production, he says the processing aspect of language is a powerful tool of assessment. Miller says that sexual selection has favored playfulness and creativity: "The attractive forms of novelty tend to rely on a uniquely human trick: the creative recombination of learned symbolic elements (e.g., words, notes, movements, visual symbols) to produce novel arrangements with new emergent meanings (e.g., stories, melodies, dances, paintings)" (413).

Miller (2000) begins his book with a detailed review of sexual selection theory and its vicissitudes. He says that mutual mate choice by females and males explains the origins of many features of human mentality, such as imagination, humor, gossip, art, music, religion, sympathy, and morality that are puzzling to social scientists that reject evolutionary explanations. He argues that these features and many others, including such apparently wasteful displays as ornate language and poetry, arose as arts of seduction. Sexual selection has favored both the production and the assessment of these features (see chapter 7 for a discussion of sexual selection).

Fitness indicators are wasteful because they operate on the handicap principle, according to which only the fittest individuals can afford to waste energy. He believes that food sharing and male parenting, which is rare among primates, arose as fitness indicators selected both by females and their offspring. Although he discusses "courtship in the Pleistocene," Miller does not refer to any specific hominins or their environments, nor does he rely on comparative data on chimpanzee behavior. He focuses almost exclusively on mate choice in sexual selection, neglecting the role of harsher forms of male competition or female control in human evolution.

In contrast, Wrangham, a chimpologist, and Peterson (Wrangham and Peterson 1996; Wrangham 1999), true to their book title, focus primarily on the role of male competition in human evolution. A major part of their book is devoted to descriptions of the raiding behavior of wild male chimpanzees, demonstrating that this behavior is not unique to our species. This is the beginning point for their effort to trace continuities in killing and warfare from ape ancestor to human ancestor. Without elaborating, they argue that these behaviors are favored by higher intelligence. In a similar vein, Richard Alexander (1989) argues that the human psyche has been shaped by selection for intelligent planning of intergroup warfare.

They describe intergroup raiding by male chimpanzees competing for dominance, territory, and access to fertile females. These are fatal raids with the apparent goal of killing males from neighboring groups. Likewise, they note that Yanomamo Indians from Brazil engage in similar raids for similar reasons. They also review data on warfare among primitive humans, showing that even relatively egalitarian hunter-gatherers regularly engage in warfare. Then they question the accuracy of popular depictions of peaceful societies like Samoa and Typee, noting that they omit violent behaviors of males.

The authors argue that hominoid males (except bonobos) are inherently violent, that is, demonic. Like human males, chimpanzee males batter and rape fertile females, as do orangutans, and chimpanzees and gorillas kill offspring of competing males when they take over their groups. Likewise, they argue that human males are violent cross-culturally, and that this is not culturally determined but shaped by sexual selection.

CONCLUSIONS ON HOMININ MENTAL EVOLUTION

The use of chimpanzee behavior as a model for the intelligence of the LCA and/or earliest hominins is a common starting point for most of these scenarios. Likewise, the belief that modern human intelligence evolved coincident with the emergence of modern humans between one hundred and sixty thousand and sixty thousand years ago is common. (Wynn's scenario is an exception to these generalizations.) Another common theme is the use of models of human intellectual development as a framework for describing increasing mental complexity (Parker and Gibson 1979; Wynn 1989; Mithen 1996). Related to this, these scenarios propose two or three stages of mental evolution. Also, two of the models argue that human mental development recapitulates the stages of human mental evolution added on the end of development in a series of ancestors (Parker and Gibson 1979; Mithen 1996). (This observation contradicts the popular notion that human mentality is a product of "neoteny" or juvenilization (Bolk 1926). See McKinney and McNamara (1991) for a discussion of heterochrony.

As indicated above, some scenarios focus primarily on the proximate mechanisms of mentality (Wynn 1989; Mithen 1996), whereas others focus primarily on the ultimate adaptive significance of mentality (Wrangham and Peterson 1996; Miller 2000). Most of those, focusing on ultimate explanations, favor the social manipulation hypothesis (Mithen 1996; Wrangham and Peterson 1996; Miller 2000), reflecting the majority view among those studying nonhuman primates. (Mithen argues that social intelligence evolved long before technological or natural history intelligence.) Wynn focuses exclusively on technology, the only hard evidence available. Two scenarios (Wrangham and Peterson 1996; Miller 2000) emphasize the role of sexual selection in human mental evolution.

The remainder of this chapter represents our attempt to put all these strands together. In many cases, scenarios are complementary rather than contradictory, emphasizing different periods, and/or different behavioral elements.

First of all, in respect to proximate mechanisms, we agree with Mithen (1996) and Parker and Gibson (1979) that there have been at least three stages of mental evolution: in Piagetian terms, these are preoperational, early concrete, and late concrete and early formal operational thinking (they could also be called associational, logical, and hypothetical thinking; see table 9.1).

According to this logic, stages of cognitive evolution from the earliest hominins until the origin of our species apparently entailed the consolidation of early and late preoperations characteristic of humans two to five years of age in *Australopithecus* and early *Homo*, the emergence of early concrete operations characteristic of children five to seven years of age in *H. erectus*, and finally the emergence of late concrete operations and the capacity for early formal operations characteristic of children eleven or twelve years of age in modern *H. sapiens*.

We also agree with Mithen (1996) that there has been an evolutionary trend from more isolated to more generalized domains of thinking. We further agree that these domains include technological and social, but also logical, domains. Moreover, we agree that they probably did not all evolve synchronously (Langer 1996). We also agree that interaction among various mental domains, whether achieved through new connections or new synchronous development, was probably critical to the emergence of human intelligence.

In contrast to Mithen (1996) and various primatologists, we believe that chimpanzees display technological and social intelligence of approximately equal magnitude. Comparative data cited above support this conclusion. Therefore, given that chimpanzees are the models for the LCA, this was probably true for early hominins as well.

This brings us to the more speculative issue of the adaptive significance of intelligence. Given the several stages of hominin evolution, we suggest that various combinations of selection pressures of changing intensities characterized different stages. Most likely these selection pressures were generated in part by chaotic changes in climate especially during glacial periods (Richerson and Boyd 2000). Specifically, in line with our preference for the extractive foraging hypothesis of subsistence (see chapter 4), we think that in the early stages of hominin evolution, natural selection favored preoperational intelligence in the physical domain for extractive foraging with tools on a wide variety of embedded food sources. And we believe kin selection favored preoperational intelligence in the social domain for symbolic communication and for imitative and teaching skills necessary for apprenticeship in extractive foraging. In this chapter we borrow concepts from linguists and developmental psychologists who study language in modern humans and great apes. These borrowed frameworks, particularly those drawn from developmental psychology, are important because they help us judge which abilities are indicative of greater or lesser linguistic complexity. This in turn can help us evaluate scenarios for the evolution of language.

Like Miller (2000) and Wrangham and Peterson (1996), we believe that sexual selection also played an important role in the evolution of human mentality. In line with our preference for the missile-throwing hypothesis for the origin of bipedalism, we think that male competition in repelling predators, competitors, and rivals occurred with low intensity from the origin of aimed throwing of missiles in basal and early hominins. As Wrangham and Peterson point out, cross-cultural and cross-species studies indicate that males use meat in courtship. Therefore, we think male competition in missile and tool technology became increasingly important as scavenging and hunting became major foraging strategies for *H. erectus* and early *H. sapiens*. We also think intragroup cooperation and intergroup competition became increasingly important in intergroup conflict during the last phase of human evolution (see chapter 10).

Likewise, we think that in the last phase of human evolution, male competition favored language and abstract reasoning for strategic planning of political and military coups against rivals inside and outside the group (see chapter 10). Finally, we agree with Miller (2000) that both male competition and female choice played an important role in shaping language, music, and art for courtship and other forms of social manipulation.

As we have seen, Darwin's argument for continuity in mentalities of apes and humans has stood the test of time. In the next chapter we will see how his ideas about the role of intergroup conflict in human evolution stand up.

10
Origins of Cultures

When two tribes of primeval man, living in the same country, came into competition, if (other circumstances being equal), the one tribe included a great number of courageous, sympathetic and faithful members, who were always ready to warn each other of danger, to aid and defend each other, this tribe would succeed better and conquer the other. Let it be borne in mind how all-important in the never-ceasing wars of savages, fidelity and courage must be. . . . A tribe rich in the above qualities would spread and be victorious over other tribes; but in the course of time it would, judging from all past history, be in its turn overcome by some other tribe still more highly endowed. Thus the social and moral qualities would tend to slowly advance and be diffused throughout the world.

—*Darwin 1871, 498*

The study of cultures is the special province of anthropologists. The anthropological sense of the term *culture* as folkways was formulated only at the end of the nineteenth and beginning of the twentieth centuries when anthropology emerged (Stocking 1968). For this reason, Darwin speaks of tribes rather than societies and cultures. Anthropologists, however, have defined culture in many different ways according to their specialties (Kroeber and Kluckhohn 1952).

Ethnographers, focusing on ideologies, rituals, and institutions of living peoples, speak of ideational cultures; human ecologists, emphasizing populations and subsistence, speak of cultural adaptations (Keesing 1974); linguists, focusing on languages, speak of cognitive cultures (Goodenough 1981); archaeologists, focusing on physical remains of past societies, speak of material cultures (Schick and Toth 1993); and primatologists, focusing on actions and products of nonverbal creatures, also speak of material cultures (McGrew 1992; Matsuzawa and Yamakoshi 1996; Van Schaik, Ancrenaz et al. 2003). In anthropological terms, cultures generally refer to the values, ideas, products, and institutions of various groups rather than the groups themselves. Typically cultures vary from location to location and through time.

As biological anthropologists in search of evolutionary continuities, we define cultures in terms that could apply to other species: "cultures are representations of knowledge socially transmitted within and between generations in groups and populations within a species that may aid them in adapting to local conditions (ecological, demographic or social)" (Parker and Russon 1996, 432). Cultural patterns are transmitted both by teaching and observational learning, including imitation (Tomasello, Kruger, et al. 1993). In other words, they are products of intelligence and, most probably, of symbolic abilities.

When and how did cultural capacities emerge and evolve in hominins? How are they related to mentality and language? Which cognitive abilities are involved in cultural transmission? How is knowledge distributed among group members? How do social relationships promote cultural transmission? What roles have intergroup contacts, including trade, raiding, and warfare, played in the evolution of cultures?

These are the questions we address in this chapter. We follow our established pattern, tracing the origins of cultures back to the last common ancestor (LCA) by examining the evidence for cultures in our closest living relatives, the chimpanzees. Then we review evidence for material cultures in hominins, which are generally cast in terms of specific tool cultures (e.g., Oldowan, Acheulean, etc.). We stop our inquiry at the dawn of the Upper Paleolithic in Europe, when most anthropological treatments of cultural evolution begin (Steward 1955; L. White 1959; Flannery 1972). Finally, we consider a few scenarios proposed to explain the nature and adaptive significance of the capacity for culture.

Engels's Labor Scenario

In his chapter "The Part Played by Labor in the Transition from Ape to Man" in his book *The Dialectics of Nature*, Engels (1896) addresses the evolutionary origins of manual dexterity, tool use, bipedal locomotion, and speech. Following Darwin, he says that the adoption of a more erect posture in walking was "*the decisive step in the transition from ape to man*," but notes that this "presupposes that in the meantime the hands became more and more devoted to other functions" (Engels 1896, 279, 280). He says that this happened to "men in the making" in the "tropical zone—probably on a great continent that has now sunk to the bottom of the Indian Ocean" (279).

Engels contrasts dexterous human hands with the undeveloped hands of great apes, which are used in knuckle walking, as well as in nest building and object dropping. He emphasizes that the labor process begins with tool making for hunting and fishing and weapons. Hunting led to the mastery of fire. He also argues that meat eating was an essential ingredient for brain growth.

A key point in his scenario is that "the hand is not only the organ of labour, *it is also the product of labour*" (Engels 1896, 281). Labor necessitated mutual support and joint activity until "men in the making arrived at the point where *they had something to say to one another*" (283). Likewise, he argues that labor led to articulate speech and thence to increased brain development: "First, comes labour, after it, and then side by side with it, articulate speech—these were the two most essential stimuli under the influence of which the brain of the ape gradually changed into that of man" (284). In other words, feedback from behavior to evolution occurred: "By the cooperation of hands, organs of speech, and brain . . . human beings became capable of executing more and more complicated operations, and of setting themselves, and achieving, higher and higher aims" (288). Engels goes on to discuss the origins of agriculture and cultural evolution.

He invokes two mechanisms for the evolution of manual dexterity. First, he argues that labor results in the inheritance of "special development of muscles, ligaments, and, over longer periods of time, bones as well, and by the ever-renewed employment of these inherited improvements" (Engels 1896, 281). Second, he argues that in the case of the hand and the rest of the organism, "the body benefited in consequence of the law of correlation of growth, as Darwin called it" (282), leading to speech and brain development, and ultimately cultural evolution.

CULTURAL MANIFESTATIONS IN HUMAN ANCESTORS

Cultural Capacities of Chimpanzees and the LCA

In the first systematic study of chimpanzee cultures in the wild, the chimpologist William McGrew (1992) used eight criteria (six of which were proposed by anthropologist Alfred E. Kroeber): innovation, dissemination, standardization, durability, diffusion, tradition, nonsubsistence, and naturalness. He notes that the only way to be certain of innovation is to see its first occurrence and its dissemination. After reviewing eight long-term studies of unprovisioned wild chimpanzees, out of thirty-two total populations of free-ranging groups, McGrew concludes that "*no single population of chimpanzees yet shows a single behavioural pattern which satisfies all eight conditions of culture. However, all conditions (except perhaps diffusion) are readily met by some chimpanzees in some cases*" (McGrew 1992, 82). In his discussion of tool use, McGrew notes that twelve populations had forty-three total habitual tool use patterns, ranging from one to eleven per population. Most of these were for acquiring or processing foods, the others for self-care or communication. Based on this, he concludes that chimpanzees do not have human culture (see table 10.1 from McGrew).

As if to meet McGrew's criteria, an apparent case of cultural dissemination from one chimpanzee population to another in West Africa was reported by Tetsuro Matsuzawa and G. Yamakoshi (1996). An adult female, Yo, born in a group that had a tradition of cracking coula nuts with stone tools, immigrated into a group without this tradition and without coula nut trees. The investigators left coula nuts and stones out and filmed young chimpanzees trying unsuccessfully to open nuts by biting them. Following this, they filmed the youngsters watching Yo open the nuts with stones. Finally, they filmed the youngsters using stones to crack open the nuts as Yo had done.

In a similar vein, Christophe Boesch (1996) describes three approaches to identifying cultural behaviors in wild chimpanzees: (1) an ecological approach controlling for ecological factors that might explain occurrences of a behavior, (2) a transmission approach that focuses on imitation and teaching, and (3) an innovation approach that focuses on evidence of social conventions after ecological factors have been ruled out. He suggests that leaf clipping—biting a leaf to pieces while drawing it back and forth across the mouth—is cultural. He bases this judgment on the changed contexts of this action from drumming display to sexual frustration at Taï. Likewise, he suggests that leaf grooming—picking motions on a leaf—is cultural. He bases this

Table 10.1. Presence of Various Types of Tool Behavior at Different Chimpanzee Study Sites

						Site			
Pattern	Gombe	Bossou	Kasoje	Taï	Kanka Sili	Assirik	Kanton, Sapo, Tiwai	Campo, Okorobiko	Kibale
Termite-fish	X								
Ant-dip	X	X	X	X		X			
Honey-dip	X			X		X			
Leaf-sponge	X	X							
Leaf-napkin	X								X
Stick-flail	X	X	X		X				
Stick-club	X	?X	X		X				
Missile-throw	X	X	X		X				
Self-tickle	X								
Play-start	X		X						
Leaf-groom	X		X						
Ant-fish		X	X						
Leaf-clip		X	X						
Gum-gouge		X							
Nut-hammer				X			X, X, X		
Marrow-pick				X					
Bee-probe				X					
Branch-haul		X							
Termite-dig								X, X	
Total	11	8	8	5	3	2	(3x)1	(2x)1	1

Source: McGrew, W. C. 1992. Chimpanzee material culture. New York: Cambridge University Press. Reprinted with the permission of Cambridge University Press.

on the changed context from an unknown function to use as a holder for ectoparasites bitten while grooming at Gombe. Boesch argues that these actions fulfill McGrew's requirement for nonsubsistence behaviors. Finally, he argues that new field data support the conclusion that chimpanzees display culture in the sense of invented and socially transmitted local traditions (also see the review article on cultural behaviors in chimpanzees [Whiten, Goodall, et al. 1999]).

Boesch's earlier work shows that wild chimpanzees display three cognitive prerequisites of culture, that is, imitation of novel movements, demonstration teaching, and symbolic communication (Boesch 1991a, 1991b). Unlike chimpanzees (and other great apes), monkeys do not imitate novel actions or teach by demonstration (Visalberghi and Fragaszy 1990, 2002). Matsuzawa and Yamakoshi's work shows another prerequisite for culture—that adolescent and adult female chimpanzees, which move between groups, can transmit cultural traditions from one group to another. Female dispersal is particularly significant for transmitting innovations given the prolonged apprenticeship in tool use chimpanzee mothers provide their offspring.

In other words, cultural transmission is facilitated by female dispersal in chimpanzee societies (Wrangham and Peterson 1996; Wrangham 2001). This pattern is found in most human societies, but absent in most monkey societies. In accord with these and other findings regarding chimpanzee demonstration teaching and imitation discussed in chapters 8 and 9, we conclude that our closest living relatives display protocultural capacities (Parker and Russon 1996). Based on our prior reasoning, we believe that at least the same level of cultural capacity was present in the last common ancestor of chimpanzees and humans.

Cultural Capacities of Early Human Ancestors

As mentioned in previous chapters, the earliest evidence of stone tool production and associated butchery and marrow consumption was 2.5 million years ago in our early ancestors *A. garhi* and *H. habilis*. We also surmised that the simple Oldowan chopper tools they produced reveal early preoperational intelligence similar to that of three- or four-year-old children, more elaborated than that of the basal hominins. Clearly, this new technology and subsistence pattern represented a significant extension of earlier extractive foraging technologies. It also suggests increased dependence on aimed throwing of missiles.

Exploitation of a rich new food source must have provided new challenges from competing scavengers and hunters and new opportunities for male competition and female choice through gifts of meat. Based on the likely social organization of the LCA, we think these hominins probably continued to live in patrilineal groups with female dispersal. Emigrating females would have continued to disperse innovative local knowledge to nearby groups. Likewise, this new pattern must have required more elaborated apprenticeships, including male gender specialization in the new activities leading to demonstration teaching by adult males. Finally, it must have entailed more elaborated communication patterns.

Cultural Capacities of Middle-Period Human Ancestors

As compared to earlier hominins, *H. erectus*, beginning 1.8 million years ago, manifested a significant increase in body and brain size and modernity, as well as a more sophisticated Acheulean tool technology and material culture. These characteristics undoubtedly helped propel these hominins out of Africa and into Asia.

Increased reliance on material culture involved in hunting and scavenging large animals, butchering meat, processing tubers, and constructing shelters must have demanded more elaborate referential communication systems for using an increased range of material culture items. Learning subsistence skills must have favored longer, more gender-specific apprenticeships than those of earlier hominins. Apprenticeships must have included practice in aimed missile throwing, recognizing signs of animal behavior, identification and transport of raw materials, and tool and shelter production, as well as tutelage in means for extracting, processing, and storing food. Presumably, as prey species became larger, related males defended rich resource caches and competed with males from adjacent groups for resources and females. Likewise, females, as well as occasional males, probably continued to transmit cultural knowledge between groups.

Cultural Capacities of Recent Human Ancestors, Archaic and Modern *Homo sapiens*

From the time of the earliest *H. sapiens*, about one hundred and sixty thousand years ago in Africa, our ancestors had achieved modern brain size. However, earlier human ancestors from the Middle Paleolithic period, beginning about two hundred and fifty thousand years ago, were already producing

Robert Bigelow's Warfare Scenario

Bigelow (1969), a New Zealand biologist, explains the threefold increase in the size of the human brain, and the associated increase in intelligence, as the consequence of selection for intergroup cooperation-for-conflict, that is, warfare. He defines warfare as "intergroup conflict with intent to kill on both sides" (57). He assumes that bigger groups depend upon increased intelligence: "Communication and efficient social organization require brains" (57).

He speaks of natural selection, but implies sexual selection in his discussion of the prevalence of polygyny and the "genetic generosity" of men in battle, and he implies group selection in his discussion of intergroup warfare. He argues that only a very strong, continuing selection pressure acting on man alone can explain the trebling of brain size in human evolution: "Warfare seems to fill all the requirements of the powerful, built-in, distinctively human force we are seeking" (Bigelow 1969, 51). He states that "*cooperation* is the secret of success in war, and cooperation requires *brains*" (15), and winners of wars produced more offspring than losers (57). Therefore, "the essence of this thesis is that the ability to learn cooperation was actually favored by the selective force of war" (19). He cites Darwin's idea that bipedalism and canine reduction were favored to free the hands for weapons, and his idea that tribal warfare selected for social and moral qualities. He faults Darwin, however, for failing to recognize the overwhelming importance of warfare.

He reviews studies of monkeys and apes, concluding that, unlike humans, chimpanzees are peaceful, and do not torture individuals captured from

---

Middle Stone Age (MSA) tools of greater range and complexity than the preceding Acheulean industry (McBrearty and Brooks 2000). As discussed in chapter 4, MSA peoples in Africa were skilled hunters and fishers using composite (hafted) projectile tools for long-distance killing, bone tools, perforated shell beads, and grindstones for grinding pigments as well as vegetable foods, behaviors usually associated with Upper Paleolithic peoples in Europe.

Their large group territories and the raw materials of their technology suggest long-distance trade networks important in risk management. These in turn suggest symbolic communication (McBrearty and Brooks 2000). The scattered, stepwise appearance of these and other advancements led McBrearty and Brooks to conclude that the transition from the MSA to the

other groups. He assumes that warfare began more than a million years ago, with australopithecines, if not sooner. He suggests that Cro-Magnon killed off Neanderthals. Much of his book is devoted to description of warfare in Mesopotamia, ancient Greece, China, and the Americas. He also discusses the military successes of the Scythians, the Mongols, and the Vikings. In addition, Bigelow reviews studies of primitive warfare in New Guinea, Australia, and Africa, noting frequent conflict over women and the greater reproductive success of successful warriors. He concludes that "most of the many factors that favor the reproductive potential of cooperative people and good warriors can be grouped under two categories: (1) the genetic effects of increased Lebensraum and (2) the genetic effects of polygamy" (Bigelow 1969, 106).

In a chapter on the argument over the peaceful versus the aggressive nature of primitive societies, Bigelow questions the distinction between instinctive or innate versus learned and intelligent behavior. He argues that human warfare, like other behaviors, is both innate and intelligent. Regarding peace, he concludes that "only an even mightier juggernaut of even more complex and all-pervading social organization can establish and maintain global law and order" (1969, 216).

He notes the biological effect of such inventions as the spear-thrower and the bow and arrow, and later, the cavalry and battleships: "Each new cultural invention in one region created a 'missile gap' that stimulated activity in other regions. The stimulus, however, was not purely cultural—it was also biological" (Bigelow 1969, 243–44). He ends his book with the question of whether we humans can change from the primitive law of cooperation-for-conflict to cooperation-for-survival.

LSA in Africa was gradual and episodic. They attribute it to intensification due to competitive responses to crowding and environmental deterioration.

This gradual pattern contrasts with the sudden "revolutionary" appearance of rapidly evolving regional Upper Paleolithic cultures in Europe associated with the immigration of anatomically modern *H. sapiens* into that region. Nevertheless, modern humans in the two regions show parallel capacities for material culture, implying symbolically mediated behaviors and intergroup alliances. These alliances were most likely marked by exchanges of mates as well as trade in local goods (Gamble 1976, 1986).

Overall, the most striking trend across the stage of human evolution is the gradually accelerating pace of cultural change leading to modern humans.

First came the static nature of culture in the last common ancestor, basal human ancestors, and the early human ancestors; second came the extremely slow pace of change in middle-period human ancestors; third came the gradual acceleration of change in archaic and even early modern *H. sapiens*; fourth came the increasing pace from the Late Stone Age in Africa and the Upper Paleolithic in Europe to the Neolithic revolution beginning ten thousand years ago, to the industrial revolution beginning in the 1700s, culminating in the blistering pace in the past century, manifested in major changes in the surface of the earth and its climate, as well as major extinctions.

CULTURE AND COOPERATION

Cooperation among kin and nonkin are prerequisites for culture. Most anthropologists would agree that fully human societies, as contrasted with most other animal societies, are based on cooperation with nonkin of the same sex and the opposite sex (Key and Aiello 1999; Knight, Dunbar, et al. 1999). In contrast, cooperation among kin is widespread among both human and nonhuman societies, owing to advantages of helping those who share genes (kin selection) (Morin, Moore, et al. 1994; Clutton-Brock 2002). Likewise, under certain conditions, cooperation among nonkin can occur through reciprocal altruism or mutual aid (Trivers 1971).

Richerson and Boyd (2003) argue that nonkin cooperation is based on innate social instincts. These include tendencies toward conformist social learning and moralistic enforcement of norms, as well as in-group identification and out-group hostility (Eibl-Eibesfeldt 1989). Richerson and Boyd argue that conformist social learning, including imitation of successful high-status models or of the most frequent behavior, biases imitation in adaptive ways. Moralistic enforcement of norms reduces within-group defection. The evolution of the capacity for morality has been a focus of interest since Darwin addressed it in *The Descent of Man* (Alexander 1987). Frans de Waal traces its origins to reciprocal altruism in the LCA (de Waal 1996).

Richerson and Boyd (2003) propose that human social instincts arose gradually during the Pleistocene epoch. They propose that more recent cultural changes have evolved through a process of *group selection on cultural variation*. They distinguish this form of selection from ordinary group selection based on genetic differences among competing groups. Noting the high rate of intermarriage among human tribal groups, they argue that group selection on cultural variants leads to extinction of cultures through incorporation into and/or emulation of the culture of the dominant group. Unlike group selection, group selection on cultural variation does not lead to physical extinction of entire competing groups. They note that coevolutionary gene-cultural processes of selection have favored genotypes better suited to live in cooperative groups. As mentioned above, Richerson and Boyd (2000, 2003) argue that cultural systems of inheritance arose in response to major unpredictable climate changes.

Echoing ideas of Darwin (1868) and Engels (1896), various anthropologists have described human cultural evolution as a positive feedback system (Childe 1951; Washburn 1960), a coevolutionary ratchet (Richerson and Boyd 2003), or an autocatalytic process leading to genetic change. Several models of culture-gene interaction have been proposed (Lumsden and Wilson 1981; Durham 1991).

The concept of *niche construction* provides useful perspective on feedback in human evolution (Odling-Smee, Laland, et al. 1996; Laland, Odling-Smee, et al. 1999; Laland, Odling-Smee, et al. 2000; Laland, Odling-Smee, et al. 2001; Odling-Smee, Laland, et al. 2003): "Niche construction refers to the activities, choices, and metabolic processes of organism, through which they define, chose, modify, and partly create their own niches. . . . In every case, however, the niche construction modifies one or more sources of natural selection in a population's environment and in doing so generates a form of feedback in evolution that is not yet fully appreciated" (Laland, Odling-Smee, et al. 2000, 132).

The novel element in the concept of niche construction comes in placing artifacts and other behavioral products outside organisms, thereby identifying two complementary systems: genetic inheritance and extragenetic ecological inheritance (Odling-Smee 1988). Niche-constructing organisms transmit modified environments, leading to modified selection pressures on succeeding generations. Although many species engage in niche-constructing behaviors including habitat selection, environmental perturbation, prediction, and nest and burrow construction, humans are particularly powerful niche constructors. Beginning in the Late Stone Age, human environmental constructions have been increasingly pervasive and long lived, as well as cumulative and progressive. Thus, human niche construction helps explain the accelerating pace of cultural evolution.

Distributed knowledge is another useful concept for understanding human cultural evolution. This is the idea that social and cultural knowledge is distributed among interacting organisms within their constructed environment, rather than being embodied in a single organism (Hutchins 1995; Strum, Forster, et al. 1997). Although other primates depend to some degree on distributed knowledge, humans do so to an unprecedented degree (Parker 2004). Language has been a powerful agent for generating and sharing distributed knowledge. Written notation systems, especially books and computers, have greatly accelerated the pace of cultural innovation, transmission, and shared creations. The breadth and magnitude of distributed knowledge has

Merlin Donald's Scenario for the Evolution of Culture, Language, and Cognition

In his book *Origins of the Modern Mind*, Merlin Donald (1991) describes three hypothetical transitions in the evolution of culture, cognition, and language:

1. from the level of *episodic culture* characteristic of apes and australopithecines to the mimetic level of culture of *H. erectus;*
2. from the *mimetic culture* characteristic of *H. erectus* to the *mythic culture* of *H. sapiens;* and
3. from the *mythic culture* of preliterate societies to the emergence of *theoretic culture* of literate societies (16–17).

Donald notes that his enterprise requires the reinterpretation of culture in cognitive terms. Likewise, he emphasizes that language is secondary to cognitive skills, and is therefore about "uniquely human styles of representation" (120).

Donald says that inquiry into cognitive evolution must begin with the study of great ape cognition and culture. Based on literature on great ape cognition and symbol learning, he concludes that these creatures display episodic culture: "Their lives are lived entirely in the present, as a series of concrete episodes" (1991, 149). He notes, however, that episodic memory is more evolved in apes than in other mammals.

He says that the earliest known hominids (now hominins), the australopithecine species, emerged about four million years ago. Based on their brain size and lack of worked stone tools, he considers that australopithecine intellect to be at the same level as that of great apes: "There is no evidence of a profound change in intellectual capacity in the reconstructed culture of the australopithecines" (Donald 1991, 105).

Donald says that *Homo erectus* was characterized by several new bodily features including increased body size and a substantial increase in brain size, culminating in a brain about 80 percent the size of modern human brains. Accordingly, he proposes that they displayed a new level of culture, mimetic culture. To explain their minds, he says we should look for a pattern of adaptation intermediate between the modern human mind and the episodic mind of the australopithecines (1991, 165). He uses the term *mimesis* to describe their dominant mode of representation: "Mimetic skill or mimesis rests on the ability to produce conscious, self-initiated, representational acts that are intentional but not linguistic" (168). Donald distinguishes among mimicry, mimesis, and imitation, though mimesis may incorporate imitation and mimicry to reenact events. He says that mimesis is intentional communication, and precedes language ontogenetically in human children.

He argues that mimetic interactions in social settings would have led to ritual dance, mimetic games, and other rituals, which would have had pedagogical functions. Stone tool making may have been the first kind of behavior that depended entirely on mimesis. Such other socially coordinated activities as cooperative hunting and fire making also relied on mimesis. Regarding language origins, Donald suggests that combined facial and vocal emotional expression may have played a primary role in social communication in mimetic culture. He believes this voluntary use of face and voice preceded their later use in language. Donald says that brain size reached its modern size with the emergence of *Homo sapiens* approximately between two hundred thousand and one hundred thousand years ago, coincident with an increasing pace of cultural change.

Regarding language evolution, Donald agrees with Lieberman and others that speech evolved in *H. sapiens* coincident with a final increase in brain size and with changes in the vocal apparatus, He differs from them, however, in his emphasis on the crucial role of cognition: "Language is, in a sense, secondary to the evolution of fundamental cognitive skills" (1991, 120). Based on the use of language in tribal societies, Donald concludes that the preeminence of myth suggests that early humans were using language for a new kind of integrative thought: "metaphorical thought could compare across episodes, deriving general principles and extracting thematic content" (215). He argues that invented symbols are the key to understanding human cognition, language, and culture. Hence, Donald concludes that the most important selection pressure on speech was a mentality that needed language to model its agendas.

Based on functions of various kinds of gestures, Donald proposes that arbitrary symbols might have been acquired through the standardization of gestures used in mimetic performances. Later he specifies several cognitive changes accompanying the evolution of the capacity for speech under cultural selection pressures including development of the speech organs, enhanced auditory perception, and new cognitive skills.

Donald argues that the narrative mode of thought is the quintessential product of language and that myth is the supreme product of the narrative mode in preliterate societies. In other words, he says, language provided a new mythic system for representing reality, which coexisted with and extended the earlier mimetic system for representing reality. He argues that whereas the emergence of mimetic and mythic cultures depended on changes in "biological hardware," the transition to theoretical culture depended upon changes in "technological hardware." *Theoretic culture* involves three new cognitive phenomena: visuographic invention, external memory, and theory construction (Donald 1991, 272). He describes three modes of visuographic innovation, or "the symbolic use of graphic devices": pictorial, ideographic, and phonological (275). Based on archaeological evidence, he says that until the emergence of cave painting in the Upper Paleolithic, visuographic invention was confined to grave and bodily decoration. They are significant because they provide a bridge to other external memory devices.

Writing developed fifteen thousand years after Upper Paleolithic cave art, but the exact break point between art and writing is obscure. He argues that the independence and arbitrariness of writing is first apparent in counting, that is, in numerical symbols. The most significant aspect of writing systems in Donald's view is that they establish what he calls "the *external memory field* or EMF, which is essentially a cognitive workspace external to biological memory" (1991, 296). Eventually, he notes, phonetic alphabetic systems emerged and spread widely owing to their greater simplicity and economy. These systems reduced the memory load associated with learning pictographs and allowed readers to map visual displays onto spoken words. Other kinds of visuosymbolic inventions include musical notation, maps, astronomical observatories, calendars, clocks, geometry, architectural plans, choreographic notations, film, and finally, computers. He emphasizes that, in addition to providing external memory fields, which can be revised and refined, these inventions allow pooling and sharing of the knowledge of many individuals. He emphasizes the unlimited size, retrievability, and modifiability of these external symbolic storage system (ESS) devices as contrasted with the limited size of biological memory systems.

increased exponentially during cultural evolution as population sizes and task specializations have increased. Exploitation of distributed knowledge generally involves cooperation with nonkin.

In the following sections we briefly describe and discuss various scenarios proposed to describe stages of evolution of culture and the selection pressures leading to the evolution of cultural capacities, and its prerequisite, cooperation among nonkin.

SCENARIOS OF CULTURAL ORIGINS

A Model for Stages of Evolution of Cultural Capacities

The most comprehensive framework for the stages of evolution of the capacity for culture, language, and cognition was proposed by the Canadian psychologist Merlin Donald (1991) in his book *Origins of the Modern Mind*. Donald describes four levels of culture: (1) episodic culture of great apes and early human ancestors, characterized by episodic memory and event representation; (2) mimetic culture of middle-period human ancestors, characterized by mimetic representation; (3) mythic culture of recent human ancestors, characterized by symbolic invention and narrative speech; and (4) theoretic culture of modern humans, characterized by extrasomatic memory storage, writing, computing, etc. (see table 10.2).

Table 10.2. Summary of Donald's (1991) Stages of Culture, Cognition, and Representation

	Culture	*Cognition*	*Representation*
Apes and australopithecines	Episodic	Episodic memory dominant device of event representation	Episodic event representation; procedural learning of signs
Homo erectus	Mimetic	Extended self-representation through mimesis	Ritual
Homo sapiens	Mythic	Symbol invention; semantic memory	Speech in narrative mode
Modern humans	Theoretic	Extrasomatic memory storage	Writing, graphing, computing

Bingham's Complete Theory of Human Evolution

In his "theory of everything," P. M. Bingham (1999) argues that human uniqueness derives from the evolution of the capacity for remote killing of conspecifics, which originated 2.5 to 2 million years ago in the form of aimed throwing and clubbing. This ability, he argues, allowed humans to transcend the limits of kin selection and cooperate with nonkin because it allowed effective punishment of defectors. Specifically, "coalitional enforcement of kinship-independent social cooperation is the fundamental thing that humans do. . . . Without exception, everything uniquely human—language, cognitive virtuosity, and so on—is either a facet of this fundamental adaptation or a subsidiary adaptation allowed by it" (249). Moreover, Bingham asserts that technical sophistication is directly proportional to the size of cooperative coalitions.

According to his "coalitional enforcement" model, the cost of punishing cheaters prevents nonhuman animals from forming nonkin coalitions. Each individual alone, cheater or punisher, has a 50 percent chance of being injured or killed when fighting tooth and claw. In contrast, "remote killing competence allows many animals to attack a target animal simultaneously. Under these conditions, the risk to individual attackers is reduced as the square of their number" (Bingham 1999, 250). If ten group members attack one cheater simultaneously, each one has only a 0.5 percent chance of injury or death.

According to Bingham, the benefits of sharing information with nonkin under coalitional reinforcement explain the evolution of language and

---

Beginning with great apes, Donald suggests that they live in the present, mentally representing only specific episodes. He proposes that australopithecines displayed the same level of culture and cognition as great apes. Based on their larger brains, he proposes that *H. erectus* displayed a level of culture based on mimetic representations of past events in rituals and other coordinated activities.

He argues that anatomically modern humans achieved the level of mythic culture based on language and characterized by narrative thought. Although this allowed them to generalize themes, they were still constrained by the limitations of biological memory. Ultimately, Donald says, humans developed theoretical culture based on visuographic invention or "symbolic use of graphic devices," which became external memory devices (275). These

> intellectual and technological virtuosity. This is because "all major increases in human adaptive sophistication will unambiguously require, and inevitably follow from, increases in coalition size" (1999, 253). He also says that human morality and guilt are predictable outcomes. He argues that expanded social cooperation can be read in the fossil record from expanded cranial volume in *H. erectus*, and secondary altriciality of dependent infants in early *Homo*.
>
> He finds anatomical evidence of throwing and clubbing in attachments for the gluteus maximus (buttocks) muscle; in this action, "the major parts of the body move in an extremely violent way. We drive forward off the back leg and then, in fierce, rapid-fire sequence, plant the front leg, rotate the hips, and torso, and shoulders followed by whipping the arms and hands. The gluteus maximus muscle contract vigorously during the violent, rapid rotations off the trunk" (Bingham 1999, 253). This muscle should be specialized immediately before the rise of *Homo*.
>
> Bingham argues that the coalitional enforcement hypothesis accounts for "the essential features of all major adaptive transitions throughout the entire two million years of human history through the present instant" (1999, 254). These include the behaviorally modern human revolution of the Upper Paleolithic, the agricultural revolution, and the rise of the modern state. Social change is driven by novel weapon technologies, which in turn drives an expanded scale of social cooperation, and so on. The atlatl and bow and arrow are examples of such technologies.

were manifested in cave paintings, astronomical observatories, maps, writing and counting systems, and other devices that produce external memory fields. These external memory fields, culminating in computers, have greatly expanded the scope of memory and facilitated pooling and sharing of knowledge.

One of Donald's main points is that each level of culture is grounded in specific cognitive skills. Likewise, he argues, each level of language is grounded in specific styles of mental representation. In other words, both culture and language are manifestations of cognition. He says very little about selection pressures favoring this complex of related abilities. In contrast, the following three scenarios for the adaptive significance of cultural capacities focus on co-operation and competition.

Scenarios of the Adaptive Nature of Culture and Social Organization

Key and Aiello (1999) argue that the highly unusual human pattern of cooperation among individuals of both the same and opposite sexes arose as an adaptation to the high energy demands of large-brained, meat-eating *H. erectus* or descendants: "We conclude that theses human patterns of cooperation are the result of changes in the energetic costs of producing large-brained offspring for males and females in association with a change to an animal based diet" (27).

In his "complete theory of everything," biologist Paul Bingham (1999) argues that the ability to form coalitions among nonkin arose out of the capacity for remote killing through aimed throwing and clubbing. He argues that remote killing facilitates coalitions among nonkin because it allows effective punishment of defectors within groups. Cooperative attacks against defectors reduce the risk of injury to the enforcers. Sharing of information among nonkin with coalitional reinforcement has driven increasing coalition size, and therefore increasing technical sophistication through the past two million years. He concludes that "without exception everything uniquely human—language, cognitive virtuosity, and so on—is either a facet of this fundamental adaptation or a subsidiary adaptation allowed by it" (Bingham 1999, 249).

New Zealand biologist Robert Bigelow (1969) proposes that the enlarged human brain and associated higher intelligence have arisen through selection for *intergroup cooperation-for-conflict*, that is, warfare. His main point is that the ability to learn to cooperate was favored by successful warfare. He reviews studies of monkeys and apes, suggesting that unlike humans, they do not kill and torture (done before reports of chimpanzee raiding and killing). He suggests that Cro-Magnon man killed Neanderthals. His review of human history and primitive tribes supports his notion of the prevalence of warfare among humans. He notes the high reproductive success of successful warriors.

Wrangham and Peterson (1996) argue that hominins elaborated themes of male bonding, violence toward females, and lethal raiding of neighboring groups derived from chimpanzee-like ancestors. They review historical and cross-cultural data on the incidence of crime and warfare to support their argument that male temperament shaped by sexual selection is the source of violence. For this reason they reject the argument that warfare is culturally determined. They trace the evolution of violence in so-called "party gang" species like chimpanzees and humans to the low cost and reproductive advantage of coalitionary bonds among males forced to live in fusion-fission groups by seasonally dispersed foods. They note that male bonding facilitates larger party sizes, which in turn are less vulnerable to attack than smaller groups.

CONCLUSIONS ABOUT THE ORIGINS OF CULTURAL CAPACITIES

Although primatologists have shown an interest in great ape cultural capacities, there are few comparative frameworks for describing levels of culture (excepting "tool cultures") among different hominin species. The most comprehensive and best known of these is Merlin Donald's (1991). We agree with his emphasis on the roles of language and intelligence in culture. Although we like his continuity approach, we think that he underestimates the mental abilities of great apes (Boysen and Berntson 1990; Parker and Gibson 1990; Parker, Mitchell, et al. 1994; Parker and McKinney 1999; Parker, Miles, et al. 1999; Russon and Begun 2004), and therefore underestimates the capacities of the basal and earliest hominins.

The adaptive significance of human cultural capacities has received relatively little attention, as compared to related abilities for language and higher intelligence. Given the close relationship among these elements, allocating scenarios among these last three chapters has been somewhat arbitrary. We have placed scenarios focusing on the evolution of cooperation in this chapter, and those focusing on language and mentality in chapters 8 and 9. Although the scenarios in this chapter do not focus on cultural capacities per se, we include them here because cooperation is prerequisite to culture. We believe that culture and language emerged through the interaction of intelligence and cooperation. Neither was sufficient by itself.

In our review, *cooperation with kin and nonkin* emerges as one of the most salient characteristics of human societies and cultures. Key and Aiello's (1999) scenario attributes kin and nonkin cooperation to selection pressures on rearing of large-brained dependent offspring. Other investigators have emphasized kin and paternal cooperation in relation to life history evolution (Lovejoy 1981; Lancaster and Lancaster 1983; O'Connell, Hawkes, et al. 1999; see chapter 6). These patterns apparently arose through kin selection and sexual selection.

In contrast, three of the adaptive scenarios we review in this chapter attribute the capacity for cooperation with nonkin to selection for coalitions against internal defectors and/or external enemies. Bingham (1999) argues that the capacity for remote killing allowed effective, low-cost enforcement of group decisions on defectors within the group. In contrast, Wrangham and Peterson (1996) and Bigelow (1969) argue that cooperation among violence-prone groups of males allowed effective raiding and warfare on other groups. The latter two scenarios focus specifically on the role of male-male competition. Taking another approach to sexual selection, Geoffrey Miller (1999, 2000) emphasizes the role of male competition and female choice in motivating such cultural productions as art, music, and science.

In our view, these scenarios are complementary, each of them focusing on different interdependent elements of human societies and cultures. One key element they neglect within nonkin cooperation is intergroup trade and marriage (Gamble 1976, 1986). Competitive feasting is another important element (Hayden 1996). Presumably, these practices emerged during the Middle and Late Stone Ages.

It is important to emphasize that nonkin cooperation alone does not generate cultures. Without higher intelligence, languages and cultures could not have emerged. Equally, language and higher intelligence could not have emerged without kin cooperation in rearing and apprenticing dependent young. Likewise, nonkin cooperate in resource defense, mating, education, politics, large-game hunting, and technology-based subsistence activities. Without this cooperation, humans could not have invented maps, writing and number systems, architecture, astronomical observatories, and other extrasomatic storage systems, which, as Donald (1991) says, allow us to transcend our limited biological memories. Without culture, humans could not create and call upon a huge reservoir of distributed knowledge, which, in turn, drives culture change. Nor, for better or worse, could humans have constructed an exponentially accelerating technological niche.

Epilogue: What about Bonobos?

For most traits discussed in this book we have used chimpanzees as models for the last common ancestor (LCA) of chimpanzees, bonobos, and humans. Given the differences between our two equally close living relatives, we should explain why we have used chimpanzees rather than bonobos in most of our reconstructions.

First, we should note that along with lesser apes, great apes share derived characteristics of shortened trunks, tail loss, and the flexible shoulder, elbow, and wrist joints underpinning suspensory locomotion (see chapter 5). All great apes except humans share ancestral characters of elongated canine teeth and thin molar enamel. They all share the derived characteristics of intelligent tool use, imitation of novel manual actions, and mirror self-recognition (see chapter 9). African apes (gorillas, chimpanzees, and bonobos) alone share the derived character of knuckle walking. Humans alone are bipedal. See table 1.

When we compare humans to the African apes, we immediately see that humans display the most divergent or uniquely derived characteristics (the focus of this book). When we compare gorillas, chimpanzees, and bonobos (pygmy chimpanzees), we see that despite many similarities, bonobos are the most behaviorally divergent of the three African apes. Moreover, bonobos and humans generally differ from other African apes behaviorally in different ways. On the other hand, bonobos are more similar to humans than chimpanzees are in a few characteristics, notably body proportions and sexual behavior.

Table 1. Shared Derived and Derived Characteristics in Apes

	Gibbons	Orangutans	Gorillas	Chimpanzees	Bonobos	Humans	
Anatomy							
Bipedality	–	–	–	–	–	+	Derived
Canine reduction	–	–	–	–	–	+	Derived
Thick molar enamel	–	–	–	–	–	+	Derived
Enlarged frontal lobes	–	+	+	+	+	+	Shared derived w/GA
Social Behaviors							
Fission-fusion group	–	+	–	+	+	+	Shared derived w/AA
Male philopatry	–	?	–	+	+	+	Shared derived w/AA
Male cooperative hunting	–	–	–	+	–	+	Shared derived w/LCA
Male cooperative defense	–	–	–	+	–	+	Shared derived w/LCA
Reproductive Behaviors							
Male dominance	–	+	+	+	–	+	Shared derived w/GA?
Pair bonding within larger group	–	–	–	–	–	+	Derived
Concealed estrus	–	–	–	–	–	+	Derived
Female orgasm	?	+	+	+	+	+	Shared derived
Juvenile provisioning	–	–	–	–	–	+	Derived
Cognitive Abilities							
Intelligent tool use	–	+	+	+	+	+	Shared derived w/GA
Stone tool production	–	–	–	–	–	+	Derived
Gestural and manual imitation	–	+	+	+	+	+	Shared derived w/GA
Communication Abilities							
Capacity for symbol use	–	+	+	+	+	+	Shared derived w/GA
Speech	–	–	–	–	–	+	Derived
Culture	–	–	–	I	–	+	Derived
Ritual	–	–	–	–	–	+	Derived
Blush (shame)	–	–	–	–	–	+	Derived
Tears	–	–	–	–	–	+	Derived
Smile	–	–	–	–	–	+	Derived

Notes: GA = great apes; AA = African apes; LCA = last common ancestor with chimpanzees and bonobos; I = incipient form

These similarities led Zihlman, Cronin, et al. (1978) to suggest that bonobos may be a better model for the last common ancestor of humans and African apes than chimpanzees. Like modern humans, bonobos display somewhat less sexual dimorphism in body size and canine length than common chimpanzees. It is important to remember, however, that they are also less dimorphic than our early ancestor *A. afarensis*. Many of the differences among African apes may be the result of scaling to body size, or allometry (Jungers and Susman 1984; Shea 1984). Such differences do not illuminate derived or ancestral relationships.

Overall, Henry McHenry argued on the basis of detailed anatomical study of African apes, australopithecines, and humans that no one living ape species has an exclusive claim to similarities to australopithecines: "The obvious conclusion from this is that the common ancestor of the African Hominoidea was not precisely like any modern hominoid and its reconstruction must derive its form from clues provided by all extinct and extant Hominoidea" (McHenry 1984, 218). (Great apes were formerly classified as Hominoidea rather than Hominidae [see chapters 2 and 3].)

In contrast to chimpanzees, bonobos live in cohesive social groups, made possible by abundant herbaceous foods. They are less aggressive than common chimpanzees. Unlike chimpanzees, female bonobos form close bonds with each other. These bonds are initiated when females change groups, and maintained by female-female sexual behavior. Bonobos of all ages engage in frequent sexual behavior, especially when eating (Chapman, White, et al. 1994; Malensky, Kuroda, et al. 1994).

Bonobos resemble humans in some aspects of their sexual behavior. Females have extended periods of sexual receptivity (and associated genital swelling), prefer frontal (ventral-ventral) copulation, possibly because of the more ventral position of their clitorises (both are juvenilized features). They also copulate during pregnancy and lactation (Blount 1990). Blount suggests that these similarities with humans arose independently in bonobos from differing selection pressures, probably for tension reduction under conditions of food competition.

It is also important to remember that sexual selection can produce rapid evolutionary changes and is likely to be involved in speciation (West-Eberhard 1983). For this reason, sexual structures and behaviors are more likely to be uniquely derived rather than shared derived characters.

In contrast to bonobos, chimpanzees are more like modern human males in their social behavior: chimpanzee males bully and harass females; like orangutans, they engage in rape. They patrol and defend territories, raiding and killing conspecifics from adjacent groups and sometimes within their own groups. Some groups also engage in collaborative hunting. Likewise, they depend upon intelligent tool use and prolonged apprenticeships to learn how to forage in the wild. Finally, chimpanzee groups show some cultural differences in behavior (see chapter 10). Therefore, we agree with Wrangham (Wrangham and Peterson 1996; Wrangham 2001) that these are probably shared derived characters, and therefore, chimpanzees are a better model for the LCA than bonobos. See also Craig Stanford's (1999) discussion of this issue.

References

Aldis, O. 1975. *Play fighting*. New York: Academic Press.

Alemseged, Z., F. Spoor, et al. 2006. A juvenile early hominin skeleton from Dikika, Ethiopia. *Nature* 443:296–301.

Alexander, R. 1987. *The biology of moral systems*. New York: Aldine de Gruyter.

Alexander, R. 1989. Evolution of the human psyche. In *The human revolution*, ed. P. Mellars and C. Stringer, 455–513. Princeton, NJ: Edinburgh University Press and Princeton University Press.

Alexander, R. D., and K. M. Noonan. 1979. Concealment of ovulation, parental care, and human social evolution. In *Evolutionary biology and human social behavior*, ed. N. A. Chagnon and W. Irons, 436–453. North Scituate, MA: Duxbury.

Alexeev, V. P. 1986. *The origin of the human race*. Moscow: Progress Publishers.

Alvarez, J. O., C. A. Lewis, et al. 1988. Chronic malnutrition, dental caries, and tooth exfoliation in Peruvian children aged 3–9 years. *American Journal of Clinical Nutrition* 48:368–372.

Alvarez, J. O., and J. M. Navia. 1989. Nutritional status, tooth eruption, and dental caries: A review. *American Journal of Clinical Nutrition* 49:417–426.

Andelman, S. J. 1987. Evolution of concealed ovulation in vervet monkeys (*Cercopithecus aethiops*). *The American Naturalist* 129:785–799.

Andrew, R. J. 1963. The origin and evolution of the calls and facial expressions of the primates. *Behaviour* 29:1–109.

Arambourg, C., and Y. Coppens. 1968. Discovery of a new australopithecine in the Omo Beds (Ethiopia). *South African Journal of Science* 64:58–59.

Arensburg, B., A. M. Tullier, et al. 1989. A middle Paleolithic human hyoid bone. *Nature* 338:758–760.

Armstrong, D., W. Stokoe, and S. E. Wilcox. 1994. Signs of the origin of syntax. *Current Anthropology* 35 (4): 349–368.

Armstrong, D. F., W. C. Stokoe, and S. E. Wilcox. 1995. *Gesture and the nature of language.* Cambridge, UK: Cambridge University Press.

Armstrong, E. 1983. Relative brain size and metabolism in mammals. *Science* 220:1302–1304.

Arnold, E. N. 1994. Investigating the origins of performance advantage: Adaptation, exaptation, and lineage effects. In *Phylogenetics and ecology*, ed. P. Eggleton and R. Vanwright, 123–168. San Diego, CA: Academic Press.

Ascenzi, A., I. Biddittu, et al. 1996. A calvarium of late *Homo erectus* from Cepraro, Italy. *Journal of Human Evolution* 31:409–423.

Asfaw, B., T. White, et al. 1999. *Australopithecus garhi*: A new species of early hominid from Ethiopia. *Science* 284:629–635.

Baker, R. R., and M. A. Bellis. 1993. Human sperm competition: Ejaculate manipulation by females and a function for the female orgasm. *Animal Behaviour* 46:887–909.

Bartholomew, G. A., and J. P. Birdsell. 1953. Ecology and the protohominids. *American Anthropologist* 12:193–214.

Bateman, A. J. 1948. Intra-sexual selection in Drosophila. *Heredity* 2:349–368.

Bates, E. 1976. *Language in context: The acquisition of pragmatics.* New York: Academic Press.

Begun, D. 1992. Miocene fossils and the chimp-human clade. *Science* 257:1929–1933.

Begun, D. R. 2004. The earliest Hominins—Is less more? *Science* 303:1478–1480.

Begun, D. R., C. V. Ward, et al., eds. 1997. *Functional phylogeny of fossils: Miocene hominid origins and adaptations.* New York: Plenum.

Bell, Q. 1977. *On human finery.* New York: Schocken Books.

Berger, T. D., and E. Trinkaus. 1993. Late archaic human traumatic injuries: Activity and or/a bias in the record (abstract). *American Journal of Physical Anthropology Supplement* 16:56.

Bermudez de Castro, J., J. Arsuaga, et al. 1997. A hominid from the lower Pleistocene of Atapuerca, Spain: Possible ancestor to Neandertals and modern humans. *Science* 276:1392–1395.

Bicchieri, M. G., ed. 1972. *Hunters and gatherers today.* Prospect Heights, IL: Waveland Press.

REFERENCES

Bickerton, D. 1981. *Roots of language*. Ann Arbor, MI: Karoma.

Bickerton, D. 1990. *Language and species*. Chicago: University of Chicago Press.

Bigelow, R. 1969. *The dawn warriors: Man's evolution toward peace*. London: The Scientific Book Club.

Bingham, P. M. 1999. Human uniqueness: A general theory. *Quarterly Review of Biology* 74 (2): 133–169.

Blackwell, L. R., and F. d'Errico. 2001. Evidence of termite foraging by Swartkrans early hominids. *Proceedings of the National Academy of Sciences* 98:1335–1337.

Blount, B. 1990. Issues in bonobo (*Pan paniscus*) sexual behavior. *American Anthropologist* 92 (3): 702–714.

Blumenschine, R. 1987. Characteristics of an early hominid scavenging niche. *Current Anthropology* 28 (4): 384–407.

Boesch, C. 1991a. Symbolic communication in wild chimpanzees. *Human Evolution* 6 (1): 81–90.

Boesch, C. 1991b. Teaching among wild chimpanzees. *Animal Behaviour* 41:530–532.

Boesch, C. 1993. Aspects of transmission of tool use in wild chimpanzees. In *Tools, language and cognition in human evolution*, ed. K. R. Gibson and T. Ingold, 171–183. Cambridge, UK: Cambridge University Press.

Boesch, C. 1994. Hunting strategies of Gombe and Taï chimpanzees. In *Chimpanzee cultures*, ed. R. Wrangham, W. C. McGrew, F. B. de Waal, and P. Heltne, 77–91. Cambridge, MA: Harvard University Press.

Boesch, C. 1996. Three approaches for assessing chimpanzee culture. In *Reaching into thought: The minds of great apes*, ed. A. Russon, K. Bard, and S. T. Parker, 404–429. Cambridge, UK: Cambridge University Press.

Boesch, C., and H. Boesch. 1989. Hunting behavior of wild chimpanzees in the Taï National Park. *American Journal of Physical Anthropology* 78:547–573.

Boesch, C., and H. Boesch-Acherman. 2000. *The chimpanzees of the Ta? forest*. Oxford, UK: Oxford University Press.

Boesch, C., P. Marchesi, et al. 1994. Is nut cracking in wild chimpanzees a cultural behaviour? *Journal of Human Evolution* 26:325–338.

Bogin, B. 1990. The evolution of human childhood: A unique growth phase and delayed maturity allow for extensive learning and complex culture. *BioScience* 40:16–25.

Bogin, B. 1999. Evolutionary perspective on human growth. *Annual Review of Anthropology* 28:109–153.

Bolk, L. 1926. On the problem of anthropogenesis. *Proc. Section Science. Kon. Akad. Wetens. Amsterdam* 29:465–475.

Bonvillian, J. D., and F. G. Patterson. 1993. Early sign language acquisition in children and gorillas: Vocabulary content and sign iconicity. *First Language* 13:315–338.

Borgia, G. 1979. Sexual selection and the evolution of mating systems. In *Sexual selection and reproductive competition in insects*, ed. M. S. Blum and N. A. Blum, 19–80. New York: Academic Press.

Bowler, M., D. Pilbeam, et al. 1982. Human origins a century after Darwin. *BioScience* 32:507–512.

Bowler, P. J. 1986. *Theories of human evolution: A century of debate, 1844–1944*. Baltimore: Johns Hopkins University Press.

Boysen, S. T., and E. J. Capaldi, eds. 1993. *The development of numerical competence: Animal and human models*. Hillsdale, NJ: Erlbaum Publishers.

Boysen, S. T., and G. Berntson. 1990. The development of numerical skills in the chimpanzee (*Pan troglodytes*). In *"Language" and intelligence in monkeys and apes*, ed. S. T. Parker and K. Gibson, 435–450. New York: Cambridge University Press.

Boysen, S. T., G. Berntson, et al. 1995. Indicating acts during counting by a chimpanzee (*Pan troglodytes*). *Journal of Comparative Psychology* 109:47–51.

Brauer, G., and E. Mbua. 1992. *Homo erectus* features used in cladistics and their variability in Asian and African hominids. *Journal of Human Evolution* 22:79–108.

Bromage, T. G., and C. Dean. 1985. Re-evaluation of the age at death of immature fossil hominids. *Nature* 167:525–527.

Brooks, D. R., and D. A. McLennan. 1991. *Phylogeny, ecology, and behavior: A research program in comparative biology*. Chicago: Chicago University Press.

Broom, R. 1938. The Pleistocene anthropoid apes of South Africa. *Nature* 142:377–379.

Brown, F. H., J. M. Harris, et al. 1985. Early *Homo erectus* skeleton from West Lake Turkana, Kenya. *Nature* 316:788–792.

Brown, J. L., and S. L. Pimm. 1985. The origin of helping: the role of variability in reproductive potential. *Journal of Theoretical Biology* 112 (3): 465–77.

Brown, R. 1973. *A first language: The early stages*. Cambridge, MA: Harvard University Press.

Brunet, M., A. Beauvilain, et al. 1996. *Australopithecus bahrelghazali*, une nouvelle especes d'hominide ancien de la region de Koro Toro [Tchad]. *Comptes Rendus de l'Academie des Sciences, Paris* 2A (322): 907–913.

Brunet, M., F. Guy, et al. 2002. A new hominid from the upper Miocene of Chad, central Africa. *Nature* 418:145–151.

Burley, N. 1979. The evolution of concealed ovulation. *American Naturalist* 114:835–858.

Buss, D. M. 1989. Sex differences in human mate preferences: Evolutionary hypotheses tested in 37 cultures. *Behavioral and Brain Sciences* 12:1–49.

Byers, S. M. 2005. *Introduction to forensic anthropology*. San Francisco: Allyn and Bacon.

Byrne, R. W. 1995. *The thinking ape: Evolutionary origins of intelligence*. Oxford, UK: Oxford University Press.

Byrne, R. 1997. Machiavellian intelligence. *Evolutionary Anthropology* 5:172–180.

Byrne, R., and A. Whiten, eds. 1988. *Machiavellian intelligence*. London: Oxford University Press.

Byrne, R., and A. Whiten. 1990. Tactical deception in primates: The 1990 database. *Primate Report* 27:1–101.

Byrne, R., and A. Whiten. 1992. Cognitive evolution in primates: Evidence from tactical deception. *Man* 27 (3): 609–627.

Calvin, W. 1983. A stone's throw and its launch window: Timing precision and its implications for language and hominid brains. *Journal of Theoretical Biology* 104:121–135.

Campbell, C. J. 2007. Primate sexuality and reproduction. In *Primates in Perspective*, ed. C. J. Campbell, A. Fuentes, K. C. MacKinnon, M. Panger, and S. K. Bearder, 423–437. New York: Oxford University Press.

Cann, R., M. Stoneking, et al. 1987. Mitochondrial DNA and human evolution. *Science* 325:31–36.

Carbonell, E., J. Bermudez de Castro, et al. 1995. Lower Pleistocene hominids and artifacts from Atapuerca-TD6 (Spain). *Science* 269:826–830.

Carbonell, E., M. Esteban, et al. 1999. The Pleistocene site of Gran Dolina, Sierra de Atapuerca, Spain: A history of the archaeological investigations. *Journal of Human Evolution* 37:313–324.

Chance, M., R. A. Mead, and A. P. Mead. 1953. Social behavior and primate evolution. *Evolution* 7:395–439.

Chapman, C. A., F. J. White, et al. 1994. Party size in chimpanzees and bonobos: A reevaluation of theory based on two similarly forested sites. In *Chimpanzee cultures*, ed. R. W. Wrangham, W. C. McGrew, F. de Waal, and P. G. Heltne, 41–58. Cambridge, MA: Harvard University Press.

Charnov, E. L., and D. Berrigan. 1993. Why do female primates have such long lifespans and so few babies? Or life in the slow lane. *Evolutionary Anthropology* 1:191–194.

Chevalier-Skolnikoff, S. 1976. The ontogeny of primate intelligence and its implications for communicative potential: A preliminary report. *Annals of the New York Academy of Sciences* 280 (Origins and evolution of language and speech): 173–211.

Childe, V. G. 1951. *Man makes himself.* New York: New American Library.

Clark, W. E. L. G. 1959. *The antecedents of man.* New York: Harper and Row.

Clark, W. E. L. G. 1964. *The fossil evidence for human evolution.* Chicago: University of Chicago Press.

Clarke, R. J., F. C. Howell, et al. 1970. More evidence of an advanced hominid at Swartkrans. *Nature* 225:1219–1222.

Clutton-Brock, T. 2002. Breeding together: Kin selection and mutualism in cooperative vertebrates. *Science* 286:69–72.

Coddington, J. A. 1988. Cladistic tests of adaptational hypotheses. *Cladistics* 4:3–22.

Conkey, M. 1980. The identification of prehistoric hunter-gatherer aggregation sites: The case of Altamira. *Current Anthropology* 21:609–629.

Conroy, G. C. 1997. *Reconstructing human origins.* New York: W. W. Norton.

Conroy, G. C., and C. J. Mahoney. 1991. Mixed-longitudinal study of dental emergence in the chimpanzee, *Pan troglodytes* (Primates, Pongidae). *American Journal of Physical Anthropology* 86:243–254.

Conroy, G. C., and M. W. Vannier. 1991. Dental development in South African australopithecines. Part II: Dental stage assessment. *American Journal of Physical Anthropology* 86:137–156.

Coon, C. 1963. *Origins of races.* London: Jonathan Cape.

Corruccini, R. S. 1978. Comparative osteometrics of the hominoid wrist joint, with special reference to knuckle-walking. *Journal of Human Evolution* 7:307–321.

Corruccini, R. S., and H. McHenry. 2001. Knuckle-walking hominid ancestors. *Journal of Human Evolution* 40:507–511.

Cronin, J. E., V. M. Sarich, et al. 1984. Molecular perspectives on the evolution of the lesser apes. In *The lesser apes,* ed. H. Preuschoft, D. J. Chivers, W. Y. Brockelman, and N. Creel, 468–485. Edinburgh: Edinburgh University Press.

Crook, J. H. 1972. Sexual selection, dimorphism, and social organization in the primates. In *Sexual selection and the descent of man, 1871–1971,* ed. B. Campbell, 231–281. Chicago: Aldine.

Curtis, G., C. Swisher, et al. 2000. *Java man: How two geologists changed the history of human evolution.* New York: Little, Brown and Co.

Dahlberg, A. A., and R. M. Menegaz-Bock. 1958. Emergence of the permanent teeth in Pima Indian children. *Journal of Dental Research* 37:1123–1140.

Dainton, M., and G. A. Macho. 1999. Did knuckle walking evolve twice? *Journal of Human Evolution* 36:171–194.

REFERENCES

Daly, M., and M. Wilson. 1983. *Sex, evolution, and behavior.* Boston: PWS Publishers.

Dart, R. 1925. Australopithecus africanus: The ape-man of South Africa. *Nature* 115:195–199.

Dart, R. 1953. The predatory transition from ape to man. *International Anthropological and Linguistic Review* 1 (14): 201–218.

Dart, R. 1959. *Adventures with the missing link.* New York: Harper & Brothers.

Darwin, C. 1859. *On the origin of species by means of natural selection, or preservation of favored races in the struggle for life.* New York: The Modern Library, Random House.

Darwin, C. 1871. *The descent of man and selection in relation to sex.* New York: The Modern Library, Random House.

Darwin, C. 1965. *The expression of the emotions in man and animals.* Chicago: University of Chicago Press (first published 1872).

Darwin, C. 1998. *The expression of emotions in man and animals.* New York: Oxford University Press. 3rd edition.

Dasen, P., and A. Heron. 1981. Cross-cultural tests of Piaget's theory. In *Handbook of cross-cultural psychology,* vol. 4, ed. H. C. Triandis and A. Heron, 295–342. Boston: Allyn and Bacon.

Dawkins, R. 1986. *The blind watchmaker.* New York: W. W. Norton.

Dawkins, R., and J. Krebs. 1978. Animal signals: Information or manipulation? In *Behavioural ecology,* ed. J. Krebs and N. B. Davies, 282–309. London: Blackwell Publishers.

Day, M. H. 1971. Postcranial remains of *Homo erectus* from Bed IV, Olduvai Gorge, Tanzania. *Nature* 232:383–387.

de Waal, F. 1983. *Chimpanzee politics.* New York: Harper and Row.

de Waal, F. 1986. The brutal elimination of a rival among captive chimpanzees. *Ethology and Sociobiology* 7:227–236.

de Waal, F. 1996. *Good natured: The origins of right and wrong in humans and other animals.* Cambridge, MA: Harvard University Press.

Dean, C. 1985. The eruption pattern of the permanent incisors and first permanent molars in Australopithecus (Paranthropus) robustus. *American Journal of Physical Anthropology* 67:251–257.

Dean, C., M. Leakey, et al. 2001. Growth processes in teeth distinguish modern humans from *Homo erectus* and earlier hominins. *Nature* 414 (6864): 628–631.

Dean, M. C. 1987. Growth layers and incremental markings in hard tissues: A review of the literature and some preliminary observations about enamel structure in *Paranthropus boisei. Journal of Human Evolution* 16:157–172.

Dixson, A. F. 1998. *Primate sexuality: Comparative studies of the prosimians, monkeys, apes, and human beings.* Oxford, UK: Oxford University Press.

Donald, M. 1991. *Origins of the modern mind.* Cambridge, MA: Harvard University Press.

Doty, R., M. Ford, et al. 1975. Changes in the intensity and pleasantness of human vaginal odor during the menstrual cycle. *Science* 190:1316–1318.

DuBois, E. 1892. Palaeontologische onderzoekingen op Java. *Verslag van het Mijnweze* third quarter:10–14.

Dunbar, R. I. M. 1993. Coevolution of neocortical size, group size and language in humans. *Behavioral and Brain Sciences* 16:681–735.

Dunbar, R. 1996. *Grooming, gossip, and the evolution of language.* Cambridge, MA: Harvard University Press.

Durham, W. 1991. *Coevolution: Genes, culture and human diversity.* Stanford, CA: Stanford University Press.

Eberhard, W. G. 1985. *Sexual selection and animal genitalia.* Cambridge, MA: Harvard University Press.

Eibl-Eibesfeldt, I. 1979. Ritual and ritualization from a biological perspective. In *Human ethology: Claims and limits of a new discipline*, ed. M. V. Cranach, K. Foppa, W. Lepenies, and D. Ploog, 3–55. Cambridge, UK: Cambridge University Press.

Eibl-Eibesfeldt, I. 1989. *Human ethology.* New York: Aldine de Gruyter.

Ekman, P. 1972. Universals and cultural differences in facial expressions of emotion. In *Nebraska Symposium on Motivation*, vol. 19, ed. J. Cole. Lincoln: University of Nebraska Press.

Elman, J., E. Bates, et al. 1996. *Rethinking innateness.* Cambridge, MA: MIT Press.

Engels, F. 1896. The part played by labor in the transition from ape to man. In *Dialectics of nature*, 279–296. Moscow: Progress Publishers.

Etkin, W. 1954. Social behavior and the evolution of man's faculties. *American Naturalist* 88:129–142.

Etler, D. A. 1996. *The fossil evidence for human evolution in Asia.* Annual Review of Anthropology 25:275–301.

Falk, D. 1992. *Braindance: New discoveries about human brain evolution.* New York: Henry Holt.

Falk, D. 2004. Prelinguistic evolution in early hominins: Whence motherese? *Behavioral and Brain Sciences* 27 (4): 491–503; discussion 503–583.

Fedigan, L. 1982. *Primate paradigms: Sex roles and social bonds.* Montreal, Quebec: Eden Press.

REFERENCES

Ferring, C. R. 1975. The Aterian in North African prehistory. In *Problems in prehistory: North Africa and the Levant*, ed. F. Wendof and A. E. Marks, 113–126. Dallas: Southern Methodist University Press.

Fisher, D. C. 1987. Mastodont procurement by Paleoindians of the Great Lakes region: Hunting or scavenging? In *The evolution of human hunting*, ed. M. H. Nitecki and D. V. Nitecki, 309–421. New York: Plenum.

Fisher, R. A. 1930. *The genetical theory of natural selection*. Oxford: Clarendon Press.

Flannery, K. 1969. Origins and ecological effects of early domestication in Iran and the Near East. In *The domestication and exploitation of plants and animals*, ed. P. J. Ucko and G. W. Dimbleby, 73–100. Chicago: Aldine.

Flannery, K. 1972. The cultural evolution of civilizations. *Annual Review of Ecology and Systematics* 3:399–426.

Fleagle, J. G. 1999. *Primate adaptation and evolution*. New York: Academic Press.

Fouts, R., D. Fouts, et al. 1984. Sign language conversational interactions between chimpanzees. *Sign Language Studies* 34:1–12.

Fouts, R., D. Fouts, et al. 1989. The infant Loulis learns signs from cross-fostered chimpanzees. In *Teaching sign language to chimpanzees*, ed. R. A. Gardner, B. T. Gardner, and T. E. V. Cantfort, 293–307. Albany: State University of New York Press.

Fox, E. A., Sitompul, A. F., et al. 1999. Intelligent tool use in wild Sumatran orangutans. In *The mentality of gorillas and orangutans*, ed. S. Parker, L. Miles, and A. Mitchell, 99–117. Cambridge: Cambridge University Press.

Fox, R. 1971. The cultural animal. In *Man and beast*, ed. J. Eisenberg, 275–292. Washington, DC: The Smithsonian Press.

Franciscus, R. G., and E. Trinkaus. 1995. Determinants of retromolar space presence in Pleistocene Homo mandibles. *Journal of Human Evolution* 28:577–795.

Gabunia, L., and A. Vekua. 1995. A Plio-Pleistocene hominid from Dmanisi, east Georgia, Caucasus. *Nature* 373:509–512.

Gabunia L., A. Vekua, et al. 2000. Earliest Pleistocene hominid cranial remains from Dmanisi, Republic of Georgia: Taxonomy, geological setting, and age. *Science* 288:1019–1025.

Gallup, G. G., Jr. 1970. Chimpanzees: Self-recognition. *Science* 167:86–87.

Gallup, G. G. 1982. Self-awareness and the emergence of mind in primates. *American Journal of Primatology* 2:237–248.

Galton, F. 1869. *Hereditary genius: An inquiry into its laws and consequences*. London: Palgrave Macmillan.

Gamble, C. 1976. Interaction and alliance in Palaeolithic society. *Man* 17:92–107.

Gamble, C. 1986. *The Palaeolithic settlement of Europe.* Cambridge, UK: Cambridge University Press.

Gangestad, S. W., and R. Thornhill. 1997. The evolutionary psychology of extra pair sex: The role of fluctuating asymmetry. *Evolution of Human Behavior* 18:69–88.

Gannon, P. J., R. L. Holloway, et al. 1998. Asymmetry of chimpanzee planum temorale: Humanlike pattern of Wernicke's brain language area homolog. *Science* 279:220–222.

Gardner, R. A., and B. T. Gardner. 1969. Teaching sign language to a chimpanzee. *Science* 165:664–672.

Gardner, R. A., B. T. Gardner, et al., eds. 1989. *Teaching sign language to chimpanzees.* Albany: State University of New York Press.

Garn, S. M., A. B. Lewis, et al. 1965a. Endocrine factors in dental development. *Journal of Dental Research* 44:243–248.

Garn, S. M., A. B. Lewis, et al. 1965b. Genetic, nutritional, and maturational correlates of dental development. *Journal of Dental Research* 44:228–242.

Ghiselin, M. 1991. Classical and molecular phylogenetics. *Boll. Zool.* 58:289–294.

Gibson, K. R., ed. 1986. *Cognition, brain size and the extraction of embedded food resources: Primate ontogeny, cognition and social behaviour.* Cambridge, UK: Cambridge University Press.

Gibson, K. R. 1988. Brain size and the evolution of language. *The genesis of language: A different judgment of evidence*, ed. M. Landsberg, 149–172. Berlin: Mouton de Gruyter.

Gibson, K. R., and T. Ingold, eds. 1993. *Tools, language and cognition in human evolution.* Cambridge, UK: Cambridge University Press.

Goldin-Meadow, S., and C. Mylander. 1998. Spontaneous sign systems created by deaf children in two cultures. *Nature* 391:279–281.

Goodall, J. 1968. Expressive movements and communication in free-ranging chimpanzees: A preliminary report. In *Primates: Studies in adaptation and variability*, ed. P. C. Jay, 313–374. New York: Holt, Rinehart and Winston.

Goodall, J. 1971. *In the shadow of man.* Boston, MA: Houghton Mifflin Company.

Goodall, J. 1986. *Chimpanzees of the Gombe.* Cambridge, MA: Harvard University Press.

Goodenough, W. 1981. *Language, culture, and society.* Menlo Park, CA: Benjamin Cummings.

Goodman, M. 1963a. Man's place in the phylogeny of primates as reflected in serum proteins. In *Classification and human evolution*, ed. S. Washburn, 204–244. Chicago: Aldine.

REFERENCES

Goodman, M. 1963b. Serological analysis of the systematics of recent hominoids. *Human Biology* 35:377–436.

Gould, S. J. 2002. *The structure of evolutionary theory.* Cambridge, MA: Harvard University Press.

Gould, S. J., and E. Vrba. 1982. Exaptation—a missing term in the science of form. *Paleobiology* 8 (1): 4–15.

Greenfield, P. M., and E. S. Savage-Rumbaugh. 1990. Grammatical combination in *Pan paniscus*: Processes of learning and invention in the evolution and development of language. In *"Language" and intelligence in monkeys and apes*, ed. S. T. Parker and K. R. Gibson, 540–548. New York: Cambridge University Press.

Groves, C. P. 1989. *A theory of human and primate evolution.* New York: Oxford University Press.

Groves, C., and V. Mazak. 1975. An approach to the taxonomy of the Hominidae: Gracile Villafrancian hominids of Africa. *Casopis pro mineralogii a geologii* 20:225–246.

Guthrie, R. D. 1976. *Body hot spots: The anatomy of human social organs and behavior.* New York: Van Nostrand Reinhold.

Haile-Selassie, Y. 2001. Late Miocene hominids from the Middle Awash, Ethiopia. *Nature* 412:178–181.

Haile-Selassie, Y., G. Suwa, et al. 2004. Late Miocene teeth from Middle Awash, Ethiopia, and early hominid dental evolution. *Science* 303:1503–1505.

Hardy, A. 1960. Was man more aquatic in the past? *New Scientist* 7 (March): 642–645.

Harrison, G. A., J. M. Tanner, et al. 1988. *Human biology: An introduction to human evolution, variation, growth, and adaptability.* 3rd ed. Oxford, UK: Oxford University Press.

Harrold, F. R. 1989. Mousterian, Chatelperronian and early Aurignacian in Western Europe: Continuity or discontinuity? In *The human revolution*, ed. P. Mellars and C. Stringer, 677–713. Princeton, NJ: Edinburgh University Press and Princeton University Press.

Hart, D., and M. P. Karmel. 1996. Self-awareness and self-knowledge in humans, apes, and monkeys. In *Reaching into thought: The minds of the great apes*, ed. A. E. Russon, K. A. Bard, and S. T. Parker, 325–357. Cambridge, UK: Cambridge University Press.

Harvey, P. H., and T. H. Clutton-Brock. 1985. Life history variation in primates. *Evolution* 39:559–581.

Harvey, P. H., and A. H. Harcourt. 1984. Sperm competition, testes size, and breeding systems in primates. In *Sperm competition and the evolution of animal mating systems*, ed. R. L. Smith, 589–600. New York: Academic Press.

Harvey, P., R. D. Martin, et al. 1987. Life histories in comparative perspective. In *Primate societies*, ed. B. Smuts, D. Cheney, R. Seyfarth, R. Wrangham, and T. Struhsaker, 181–196. Chicago: University of Chicago Press.

Harvey, P. H., and M. D. Pagel. 1991. *The comparative method in evolutionary biology*. Oxford Series in Ecology and Evolution. Oxford: Oxford University Press.

Harvey, P. H., and A. F. Read. 1988. How and why do mammalian life histories vary? In *Evolution of life histories of mammals: Theory and pattern*, ed. M. S. Boyce, 213–232. New Haven, CT: Yale University Press.

Hawkes, K. 2003. Grandmothers and the evolution of human longevity. *American Journal of Human Biology* 15:380–400.

Hawkes, K. 2004. The grandmother effect. *Nature* 428:128–129.

Hawkes, K., J. F. O'Connell, et al. 1998. Grandmothering, menopause, and the evolution of life histories. *Proceedings of the National Academy of Sciences of the USA* 95:1336–1339.

Hayden, B. 1994. Competition, labor, and complex hunter-gatherers. In *Key issues in hunter-gatherer research*, ed. J. E. S. Burch and L. J. Ellanna, 223–239. Oxford: Berg Publishers, Inc.

Hayden, B. 1996. Feasting in prehistoric and traditional societies. In *Food and the status quest*, ed. P. Wiessner and W. Schiefenhovel, 127–146. Providence, RI: Berghan Books.

Hayes, K. J., and C. Hayes. 1951. Imitation in a home-raised chimpanzee. *Journal of Comparative Physiological Psychology* 45:450–459.

Heinzelin, J., J. D. Clark, et al. 1999. Environment and behavior of 2.5 million-year-old Bouri hominids. *Science* 284:625–629.

Heistermann, M., T. Ziegler, et al. 2001. Loss of oestrus, concealed ovulation, and paternity confusion in free-ranging Hanuman langurs. *Proceedings of the Royal Society of London: B* 268:2445–2451.

Helm, S. 1969. Secular trend in tooth eruption: A comparative study of Danish school children of 1913 and 1965. *Archaeology of Oral Biology* 14:1177–1191.

Hennig, W. 1979. *Phylogenetic systematics*. Urbana: University of Illinois Press.

Hewes, G. 1961. Food transport and the origin of Hominid bipedalism. *American Anthropologist* 63:687–710.

Hewes, G. 1964. Hominid bipedalism: Independent evidence for the food-carrying theory. *Science* 146 (4): 6–18.

Hewes, G. 1973. Primate communication and the origin of language. *Current Anthropology* 14:5–24.

Heyes, C., and B. Galef, eds. 1996. *Social learning in animals*. New York: Academic Press.

Hockett, C., and S. Ascher. 1964. The human revolution. *Current Anthropology* 5:135–168.

REFERENCES

Hoffman, M. A. 1983. Energy metabolism, brain size and longevity in mammals. *Quarterly Review of Biology* 58:495–512.

Hohn, A. A., M. D. Scott, et al. 1989. Growth layers in teeth from known-age free ranging bottlenose dolphins. *Marine Mammal Science* 5:315–342.

Holloway, R. L. 1985. The poor brain of *Homo sapiens neanderthalensis*, see what you please... In *Ancestors: The hard evidence*, ed. E. Delson, 319–324. New York: Alan R. Liss.

Houde, A. E. 2001. Sex roles, ornaments, and evolutionary explanation. *Proceedings of the National Academy of Sciences of the USA* 98:12857–12859.

Hrdlicka, A. 1927. The Neanderthal phase of man. *The Journal of the Royal Anthropological Institute of Great Britain and Ireland* 57:249–273.

Hrdy, S. B. 1977. *The langurs of Abu: Female and male strategies of reproduction.* Cambridge, MA: Harvard University Press.

Hrdy, S. B. 1979. Infanticide among animals: A review, classification, and examination of the implications for the reproductive strategies of females. *Ethology and Sociobiology* 1:13–40.

Hrdy, S. 1981. *The woman that never evolved.* Cambridge, MA: Harvard University Press.

Hrdy, S. 1999. *Mother nature.* New York: Random House.

Humphrey, N. K. 1976. The social function of intellect. In *Growing points in ethology*, ed. P. P. G. Bateson and R. A. Hinde, 303–317. Cambridge, UK: Cambridge University Press.

Hunt, K. 1994. The evolution of human bipedality: Ecology and functional morphology. *Journal of Human Evolution* 26:183–202.

Hurme, V. O. 1949. Ranges of normalcy in the eruption of permanent teeth. *Journal of Dentistry for Children* 16:11–15.

Hutchins, E. 1995. *Cognition in the wild.* Cambridge, MA: MIT Press.

Huxley, J. 1942. *Evolution: The modern synthesis.* London: Allen and Unwin.

Huxley, T. H. 1959. *Man's place in nature.* Ann Arbor: University of Michigan Press.

Inhelder, B., and J. Piaget. 1958. *The growth of logical thinking from childhood to adolescence.* New York: Basic Books.

Isaac, G. 1980. Casting the net wide: A review of archaeological evidence for early hominid land use and ecological relations. In *Current argument on early man*, ed. L. K. Konigsson, 226–251. Oxford: Pergamon.

Jablonski, N., and G. Chaplin. 1993. Origin of habitual terrestrial bipedalism in the ancestor of the Hominidae. *Journal of Human Evolution* 24:259–280.

Jasieńska, G., A. Ziomkiewicz, et al. 2004. Large breast and narrow waists indicate high reproductive potential in women. *Proceedings of the Royal Society of London: Series B* 271:1213–1217.

Jensvold, M. L., and R. A. Gardner. 2000. Interactive use of sign language by cross-fostered chimpanzees (*Pan troglodytes*). *Journal of Comparative Psychology* 114 (4): 335–346.

Johanson, D. 1986. A new partial skeleton of *Homo habilis* from Olduvai Gorge, Tanzania. *Nature* 327:205–209.

Johanson, D., T. White, et al. 1978. A new species of the genus *Australopithecus* (Primates: Hominidae) from the Pliocene of eastern Africa. *Kirtlandia* 28:1–14.

Jolly, A. 1966. Lemur social behaviour and primate intelligence. *Science* 153:501–506.

Jolly, C. 1972. The seed eaters: A new model for hominid differentiation based on a baboon analogy. *Man* 5:5–27.

Jungers, W. L. 1982. Lucy's limbs: Skeletal allometry and locomotion in *Australopithecus afarensis*. *Nature* 297:676–678.

Jungers, W. L. 1985. Body size and scaling of limb proportions in primates. In *Size and scaling in primate biology*, ed. W. L. Jungers, 345–381. New York: Plenum.

Jungers, W., and R. L. Susman. 1984. Body size and skeletal allometry in African apes. In *The pygmy chimpanzee*, ed. R. L. Susman, 131–177. New York: Plenum.

Keesing, R. 1974. Theories of Culture. *Annual Reviews of Anthropology* 3:73–97.

Key, C. A., and L. C. Aiello. 1999. The evolution of social organization. In *The evolution of culture*, ed. C. Knight, R. Dunbar, and C. Power, 15–33. New Brunswick, NJ: Rutgers University Press.

King, W. 1864. The reputed fossil man of the Neanderthal. *Quarterly Journal of Science* 1:88–97.

Klein, R. G. 1987. Problems and prospects in understanding how early people exploited animals. In *The evolution of human hunting*, ed. M. H. Nitecki and D. V. Nitecki, 11–45. New York: Plenum.

Klein, R. G. 1999. *The human career: Human biological and cultural origins*. Chicago: University of Chicago Press.

Knight, C., R. I. M. Dunbar, et al. 1999. An evolutionary approach to human culture. In *The evolution of culture*, ed. C. Knight, R. I. M. Dunbar, and C. Power, 1–14. New Brunswick, NJ: Rutgers University Press.

Kohler, W. 1927. *The mentality of apes*. New York: Vintage.

Kordos, L., and D. R. Begun. 2002. A late Miocene subtropical swamp deposit with evidence of the origin of African apes and humans. *Evolutionary Anthropology* 11 (2): 45–57.

Kortlandt, A., and M. Kooij. 1963. Protohominid behavior in primates. *Symposium of the Zoological Society, London* 10:61–88.

Kramer, A. 1993. Human taxonomic diversity in the Pleistocene: Does *Homo erectus* represent multiple hominid species? *American Journal of Physical Anthropology* 91:161–171.

Krings, M., A. Stone, et al. 1997. Neanderthal DNA sequences and the origin of modern humans. *Cell* 90:19–30.

Kroeber, A. E., and C. Kluckhohn. 1952. *Culture*. Cambridge, MA: Harvard University Press.

Kuhlmeier, V. A., S. T. Boysen, et al. 1999. Scale-model comprehension by chimpanzees (*Pan troglodytes*). *Journal of Comparative Psychology* 113 (4): 396–402.

Laland, K. N., F. J. Odling-Smee, et al. 1999. Evolutionary consequences of niche construction and their implications for ecology. *Proc Natl Acad Sci USA* 96 (18): 10242–10247.

Laland, K., J. Odling-Smee, et al. 2000. Niche construction, biological evolution and cultural change. *Behavioral and Brain Sciences* 23 (1): 131–146.

Laland, K., J. Odling-Smee, et al. 2001. Niche construction, ecological inheritance, and cycles of contingency in evolution. In *Cycles of contingency: Developmental systems and evolution*, ed. S. Oyama, P. Griffiths, and R. Gray, 117–126. Cambridge, MA: MIT Press.

Lancaster, J. 1978. Carrying and sharing in human evolution. *Human Nature* 1 (2): 82–89.

Lancaster, J. B. 1986. Human adolescence and reproduction. In *School-age pregnancy and parenthood: Biosocial dimensions*, ed. J. B. Lancaster and B. A. Hamburg, 17–37. Chicago: Aldine.

Lancaster, J. B., and B. J. King. 1985. An evolutionary perspective on menopause. In *In her prime*, ed. J. K. Browk and K. Kerns, 13–20. South Hadley, MA: Bergin and Garvey.

Lancaster, J. B., and C. S. Lancaster. 1983. Parental investment: The hominid adaptation. In *How humans adapt: A biocultural odyssey*, ed. S. Ortner, 33–65. Washington, DC: Smithsonian.

Landau, M. 1991. *Narratives of human evolution*. New Haven, CT: Yale University Press.

Langer, J. 1993. Comparative cognitive development. In *Tools, language and cognition in human evolution*, ed. K. R. Gibson and T. Ingold, 300–313. Cambridge, UK: Cambridge University Press.

Langer, J. 1996. Heterochrony and the evolution of primate cognitive development. In *Reaching into thought: The minds of great apes*, ed. A. Russon, K. Bard, and S. T. Parker, 257–277. Cambridge, UK: Cambridge University Press.

Langer, J. 2000a. The descent of cognitive development. *Developmental Science* 3 (4): 361–379 and 385–389.

Langer, J. 2000b. The heterochronic evolution of primate cognitive development. In *Brains, bodies, and behavior: The evolution of human development*, ed. S. T. Parker, J. Langer, and M. McKinney, 215–236. Santa Fe, NM: School of American Research.

Leakey, L. S. B. 1959. A new fossil skull from Olduvai. *Nature* 184:491–493.

Leakey, L. S. B., P. V. Tobias, et al. 1964. A new species of the genus *Homo* from Olduvai Gorge. *Nature* 202:7–9.

Leakey, M. G., C. S. Feibel, et al. 1995. New four million year old hominid species from Kanapoi and Allia Bay, Kenya. *Nature* 376:565–571.

Leakey, M. G., C. S. Feibel, et al. 1998. New specimens and confirmation of an early age for *Australopithecus anamensis*. *Nature* 393 (6680): 62–66.

Leakey, M., and R. Hay. 1979. Pliocene footprints in Laetoli Beds at Laetoli, Northern Tanzania. *Nature* 278:383.

Leakey, M., F. Spoor, et al. 2001. New hominin genus from eastern Africa shows diverse middle Pliocene lineages. *Nature* 410:433–440.

Leakey, R. 1973. Evidence for an advanced Plio-Pliestocene Hominid from East Rudolf, Kenya. *Nature* 42:447–450.

Leakey, R. E. F., and A. Walker. 1985a. A fossil skeleton 1,600,000 years old: *Homo erectus* unearthed. *National Geographic* 168:625–629.

Leakey, R. E. F., and A. Walker. 1985b. Further hominids from the Plio-Pleistocene of Koobi Fora, Kenya. *American Journal of Physical Anthropology* 67:135–163.

Lee, R. B. 1979. *The !Kung San: Men, women and work in a foraging society*. New York: Cambridge University Press.

Lee-Thorp, J., and M. Sponheimer. 2007. Contributions of stable light isotopes to paleoenvironmental reconstruction. In *Handbook of paleoanthropology*, ed. T. Hardt, W. Henke, and I. Tattersall, 289–310. Berlin: Springer Verlag.

Leigh, S. R., and G. E. Blomquist. 2007. Life history. In *Primates in Perspective*, ed. C. J. Campbell, A. Fuentes, K. C. MacKinnon, M. Panger, and S. K. Bearder, 396–407. New York: Oxford University Press.

Lewis, A. B., and S. M. Garn. 1960. The relationship between tooth formation and other maturational factors. *The Angle Orthodontist* 30:70–77.

Lewis-Williams, D. 2002. *The mind in the cave*. London: Thames and Hudson.

Liebermann, D., and R. McCarthy. 1999. The ontogeny of cranial base angulation in humans and chimpanzees and its implications for reconstructing pharyngeal dimensions. *Journal of Human Evolution* 36:487–517.

Lillegraven, J. A., S. D. Thompson, et al. 1967. The origin of eutherian mammals. *Biological Journal of the Linnaean Society* 32:281–336.

Linnaeus, C. 1758. *Systema naturæ per regna tria naturæ, secundum classes, ordines, genera, species, cum characteribus, differentiis, synonymis, locis*. Tomus I. Editio decima, reformata. Holmiæ: Laurentii Salvii.

REFERENCES

Linton, S. 1971. Woman the gatherer: Male bias in anthropology. In *Women in perspective: A preliminary source book*, ed. S. E. Jacobs, 9–21. Urbana: University of Illinois Press.

Livingstone, F. B. 1958. Anthropological implications of sickle cell gene distribution in West Africa. *American Anthropologist* 60 (3): 533–562.

Livingstone, F. B. 1972. Did the australopithecines sing? *Current Anthropology* 14 (1–2): 25–29.

Lovejoy, C. O. 1981. The origin of man. *Science* 211:341–350.

Lovejoy, C. O., K. G. Heiple, et al. 2001. Palaeoanthropology: Did our ancestors knuckle-walk? *Nature* 410 (6826): 325–326.

Lumsden, C., and E. O. Wilson. 1981. *Genes, mind, and culture*. Cambridge, MA: Harvard University Press.

MacLarnon, A. M., and G. P. Hewitt. 1999. The evolution of human speech: The role of enhanced breathing control. *American Journal of Physical Anthropology* 109:341–363.

Malensky, R. K., S. Kuroda, et al. 1994. The significance of terrestrial herbaceous foods for bonobos, chimpanzees, and gorillas. In *Chimpanzee cultures*, ed. R. W. Wrangham, W. C. McGrew, F. de Waal, and P. G. Heltne, 59–76. Cambridge, MA: Harvard University Press.

Malina, R. M., and C. Bouchard. 1991. *Growth, maturation, and physical activity*. Champaign, IL: Human Kinetics.

Mann, A. E. 1975. Some paleodemographic aspects of the South African australopithecines. In *University of Pennsylvania publications in anthropology*. Philadelphia: University of Pennsylvania.

Manzi, G., F. Mallegni, et al. 2001. A cranium for the earliest Europeans: Phylogenetic position of the hominid from Ceprano, Italy. *Proceedings of the National Academy of Sciences* 98:10011–10016.

Marlowe, F., C. Apicella, et al. 2005. Men's preferences for women's profile waist-to-hip ratio in two societies. *Evolution and Human Behavior* 26:458–468.

Marlowe, F., and A. Wetsman. 2001. Preferred waist-to-hip ratio and ecology. *Personality and Individual Differences* 30:481–489.

Martin, A. 1998. Origin of semantic knowledge and the origin of words in the brain. In *The origin and diversification of language*, ed. N. Jablonski and L. Aiello, 69–88. San Francisco: California Academy of Science.

Martin, R. D. 1981. Relative brain size and basal metabolic rate in terrestrial vertebrates. *Nature* 293:57–60.

Martin, R. D. 1983. *Human brain evolution in ecological context*. New York: American Museum of Natural History.

Marzke, M. W. 1986. Tool-use and the evolution of hominid hands and bipedality. In *Primate evolution*, ed. J. G. Else and P. C. Lee, 203–209. Cambridge, UK: Cambridge University Press.

Marzke, M. W., J. M. Longhill, et al. 1988. Gluteus maximus muscle function and the origin of hominid bipedality. *Am J Phys Anthropology* 77 (4): 519–528.

Mason, O. T. 1966 (1895). *The origins of invention.* Cambridge, MA: MIT Press.

Matsuzawa, T., and G. Yamakoshi. 1996. Comparison of chimpanzee material culture between Bossou and Nimba, West Africa. In *Reaching into thought: The minds of the great apes,* ed. A. E. Russon, K. Bard, and S. T. Parker, 211–232. Cambridge, UK: Cambridge University Press.

Matsuzawa, T., ed. 2001. *Primate origins of cognition and behavior.* Tokyo: Springer.

Mayr, E. 1950. Taxonomic categories in fossil hominids. *Cold Spring Harbor Symposium on Quantitative Biology* 15:109–118.

Mayr, E., and W. Provine, eds. 1989. *The evolutionary synthesis.* Cambridge, MA: Harvard University Press.

McBrearty, S., and A. Brooks. 2000. The revolution that wasn't: A new interpretation of the origin of modern human behavior. *Journal of Human Evolution* 39:453–563.

McGrew, W. 1992. *Chimpanzee material culture.* New York: Cambridge University Press.

McGrew, W. C. 1999. Commentary on "The raw and the stolen." *Current Anthropology* 40:582–583.

McHenry, H. 1984. The common ancestor: A study of the postcranium of *Pan paniscus, Australopithecus* and other hominoids. In *The pygmy chimpanzee,* ed. R. L. Susman, 201–230. New York: Plenum.

McHenry, H. 1986. The first bipeds: A comparison of the *A. afarensis* and *A. africanus* postcranium and implications for the evolution of bipedalism. *Journal of Human Evolution* 15:177–191.

McHenry, H. M. 1988. New estimates of body weight in early hominids and their significance to encephalization and megadontia in "robust" australopithecines. In *The evolutionary history of the "robust" australopithecines,* ed. F. E. Grine, 133–146. New York: Aldine de Gruyter.

McHenry, H. 1992. Body size and proportion in the early hominids. *American Journal of Physical Anthropology* 86:407–431.

McHenry, H. 1996. Sexual dimorphism in fossil hominids and its socioecological implications. *The archaeology of human ancestry,* ed. J. Steele and S. Shennan, 91–103. London: Routledge.

McHenry, H., and L. R. Berger. 1998. Body proportions in *Australopithecus afarensis* and *A. africanus* and the origin of the genus *Homo. Journal of Human Evolution* 35:1–22.

McKinney, M., and K. McNamara. 1991. *Heterochrony: The evolution of ontogeny.* New York: Plenum.

REFERENCES

Mellars, P. 1998. Neanderthals, modern humans and the archaeological evidence for language. In *The origin and diversification of language*, ed. N. Jablonski and L. Aiello, 89–115. San Francisco: California Academy of Sciences.

Mellars, P., and C. Stringer, eds. 1989. *The human revolution*. Princeton, NJ: Princeton University Press.

Mercader, J., M. Panger, et al. 2002. Excavation of a chimpanzee stone tool site in the African rainforest. *Science* 296 (5572): 1452–1455.

Miles, H. L. 1990. The cognitive foundations for reference in a signing orangutan. In *"Language" and intelligence in monkeys and apes*, ed. S. T. Parker and K. R. Gibson, 511–539. New York: Cambridge University Press.

Miles, H. L. 1994. Me Chantek: The development of self-awareness in a signing gorilla. In *Self-awareness in animals and humans*, ed. S. T. Parker, R. W. Mitchell, and M. L. Boccia, 254–272. New York: Cambridge University Press.

Miles, H. L., R. Mitchell, et al. 1996. Simon says: The development of imitation in an enculturated orangutan. In *Reaching into thought: The minds of great apes*, ed. A. Russon, K. Bard, and S. T. Parker, 278–299. Cambridge, UK: Cambridge University Press.

Millar, J. S. 1977. Adaptive features of mammalian reproduction. *Evolution* 31:370–386.

Millar, J. S., and R. M. Zammuto. 1983. Life histories of mammals: An analysis of life tables. *Ecology* 64:631–635.

Miller, G. 1999. Sexual selection for cultural displays. In *The evolution of culture*, ed. R. Dunbar, C. Knight, and C. Power, 71–90. New Brunswick, NJ: Rutgers University Press.

Miller, G. 2000. *The mating mind: How sexual choice shaped the evolution of human nature.* New York: Anchor Books, Random House.

Miller, G. F. 2003. Fear of fitness indicators: How to deal with our ideological anxieties about the role of sexual selection in the origins of human culture. In *Being human: Science, culture, and fear*, 65–79. Wellington, NZ: Proceedings of a conference sponsored by the Royal Society of New Zealand, series 63, Royal Society of New Zealand.

Milner, R. 1994. *The encyclopedia of evolution: Humanity's search for its origins.* New York: Henry Holt.

Milton, K. 1988. Foraging behaviour and the evolution of primate intelligence. In *Machiavellian intelligence*, ed. R. Byrne and A. Whiten, 285–305. Oxford, UK: Oxford University Press.

Minugh-Purvis, N., J. Radovcic, et al. 2000. Krapina 1: A juvenile Neanderthal from the early late Pleistocene of Croatia. *American Journal of Physical Anthropology* 11 (3): 393–424.

Mitchell, R. W. 1986. A framework for discussing deception. In *Deception: Perspectives on human and nonhuman deceit*, ed. R. W. Mitchell and N. Thompson, 3–40. Albany: State University of New York Press.

Mitchell, R. W. 1997. Kinesthetic-visual matching and the self concept as explanations of mirror-self recognition. *Journal for the Theory of Social Behaviour* 27 (1): 17–39.

Mitchell, R. W. 1999. Scientific and popular conceptions of the psychology of great apes from the 1700's to the 1970's: Deja vu all over again. *Primate Report* 53:3–118.

Mithen, S. 1993. Tool-use and language in apes and humans. *Cambridge Archeological Journal* 3 (2): 285–300.

Mithen, S. 1996. *The prehistory of the mind.* New York: Thames and Hudson.

Møller, A. P., and R. Thornhill. 1998. Bilateral symmetry and sexual selection: A meta-analysis. *American Naturalist* 151:174–192.

Moorrees, C. F. A., and R. L. Kent. 1978. A step function model using tooth counts to assess the developmental timing of the dentition. *Annals of Human Biology* 5:55–68.

Morgan, E. 1982. *The aquatic ape.* London: Souvenir.

Morin, P. A., J. J. Moore, et al. 1994. Kin selection, social structure, gene flow, and the evolution of chimpanzees. *Science* 265:1193–1201.

Morris, D. 1967. *The naked ape.* New York: McGraw-Hill.

Nelson, K. 1996. *Language in cognitive development: The emergence of the mediated mind.* New York: Cambridge University Press.

Nettle, D. 1999. Language variation and the evolution of societies. In *The evolution of culture,* ed. R. Dunbar, C. Knight, and C. Power, 214–227. New Brunswick, NJ: Rutgers University Press.

Nichols, J. 1998. The origins and dispersal of languages: Linguistic evidence. In *The origins and dispersal of language,* ed. N. Jablonski and L. Aiello, 127–170. San Francisco: California Academy of Sciences.

Nishida, T., T. Kano, et al. 1999. Ethogram and ethnography of Mahale chimpanzees. *Anthropological Science* 107:141–188.

Nissen, H. W., and A. H. Riesen. 1964. The eruption of the permanent dentition of chimpanzee. *American Journal of Physical Anthropology* 22:285–294.

Niswander, J. D., and C. Sujaku. 1965. Permanent tooth eruption in children with major physical defect and disease. *Journal of Dentistry for Children* 32:266–268.

Noble, W., and I. Davidson. 1996. *Human evolution, language and mind.* Cambridge, UK: Cambridge University Press.

O'Connell, J., K. Hawkes, et al. 1988. Hazda scavenging: Implications for Plio-Pleistocene hominid subsistence. *Current Anthropology* 29:356–363.

O'Connell, J. F., K. Hawkes, et al. 1999. Grandmothering and the evolution of *Homo erectus. Journal of Human Evolution* 36:461–485.

REFERENCES

Odling-Smee, J. 1988. Niche-constructing phenotypes. In *The role of behavior in evolution*, ed. H. C. Plotkin, 73–132. Cambridge, MA: MIT Press.

Odling-Smee, J., K. Laland, et al. 1996. Niche construction. *American Naturalist* 147 (4): 641–648.

Odling-Smee, J., K. Laland, et al. 2003. *Niche construction: The neglected process in evolution.* Princeton, NJ: Princeton University Press.

O'Sullivan, C., and C. P. Yeager. 1989. Communicative context and linguistic competence: The effects of social setting on a chimpanzee's conversational skill. In *Teaching sign language to chimpanzees*, ed. R. A. Gardner, B. Gardner, and T. Van Cantfort, 269–279. Albany: State University of New York Press.

Oswalt, W. 1973. *Habitat and technology.* New York: Holt, Rinehart & Winston.

Pagel, M., and W. Bodmer. 2005. A naked ape would have fewer parasites. *Proceedings of the Royal Society of London: B (Supplement)* 270:S117–S119.

Panger, M., A. Brooks, et al. 2002. Older than the Oldowan? Rethinking the emergence of hominin tool use. *Evolutionary Anthropology* 11:235–245.

Parker, S. T. 1984. Playing for keeps: An evolutionary perspective on human games. In *Play in animals and humans*, ed. P. K. Smith, 271–293. London: Blackwell.

Parker, S. T. 1985. A social technological model for the evolution of language. *Current Anthropology* 26 (5): 617–639.

Parker, S. T. 1987. A sexual selection model for hominid evolution. *Human Evolution* 2 (3): 235–253.

Parker, S. T. 1990. The origins of comparative developmental evolutionary studies of primate mental abilities. In *"Language" and intelligence in monkeys and apes*, ed. S. T. Parker and K. R. Gibson, 3–64. New York: Cambridge University Press.

Parker, S. T. 1996. Apprenticeship in tool-mediated extractive foraging: The origins of imitation, teaching, and self-awareness in great apes. In *Reaching into thought: The minds of great apes*, ed. A. Russon, K. Bard, and S. T. Parker, 348–370. Cambridge, UK: Cambridge University Press.

Parker, S. T. 2000. *Homo erectus*: A turning point in human evolution. In *Brains, bodies, and behavior: The evolution of human development*, ed. S. T. Parker, J. Langer, and M. McKinney, 279–318. Santa Fe, NM: School of American Research Press.

Parker, S. T. 2004. The cognitive complexity of social organization and socialization in wild baboons and chimpanzees: Guided participation, socializing interactions, and event representation. In *The evolution of thought*, ed. A. Russon and D. Begun, 45–60. Cambridge, UK: Cambridge University Press.

Parker, S. T., and K. R. Gibson. 1977. Object manipulation, tool use, and sensorimotor intelligence as feeding adaptations in cebus monkeys and great apes. *Journal of Human Evolution* 6:623–641.

Parker, S. T., and K. R. Gibson. 1979. A developmental model for the evolution of language and intelligence in early hominids. *Behavioral and Brain Sciences* 2:367–408.

Parker, S. T., and K. R. Gibson, eds. 1990. *"Language" and intelligence in monkeys and apes.* New York: Cambridge University Press.

Parker, S. T., J. Langer, et al., eds. 2005. *Biology and knowledge revisited: From neurogenesis to psychogenesis.* Mahwah, NJ: Lawrence Erlbaum Associates.

Parker, S. T., and M. L. McKinney. 1999. *Origins of intelligence: The evolution of cognitive development in monkeys, apes, and humans.* Baltimore: Johns Hopkins University Press.

Parker, S. T., and C. Milbrath. 1993. Higher intelligence, propositional language, and culture as adaptations for planning. In *Tools, language, and cognition in human evolution*, ed. K. R. Gibson and T. Ingold, 314–333. Cambridge: Cambridge University Press.

Parker, S. T., H. L. Miles, et al., eds. 1999. *The mentalities of gorillas and orangutans.* Cambridge, UK: Cambridge University Press.

Parker, S. T., R. W. Mitchell, et al., eds. 1994. *Self-awareness in animals and humans.* New York: Cambridge University Press.

Parker, S. T., and A. Russon. 1996. On the wild side of culture and cognition in the great apes. In *Reaching into thought*, ed. A. Russon, K. Bard, and S. T. Parker, 430–450. Cambridge: Cambridge University Press.

Patterson, F. 1980. Innovative use of language by gorilla: A case study. In *Children's Language*, vol. 2, ed. K. Nelson, 497–561. New York: Gardner.

Patterson, F., and R. Cohn. 1994. Self-recognition and self-awareness in lowland gorillas. In *Self-awareness in animals and humans*, ed. S. T. Parker, R. W. Mitchell, and M. L. Boccia, 273–290. New York: Cambridge University Press.

Pawlowski, B. 1999. Loss of oestrus and concealed ovulation in human evolution: The case against the sexual selection hypothesis. *Current Anthropology* 40:257–275.

Piaget, J., and B. Inhelder. 1967. *The child's conception of space.* New York: W. W. Norton.

Piaget, J., and B. Inhelder. 1969. *The psychology of the child.* New York: Basic Books.

Pianka, E. R. 1970. On r and K selection. *American Naturalist* 104:592–597.

Pickford, M., and B. Senut. 2001. "Millenium ancestor," a 6-million-year-old bipedal hominid from Kenya. *South African Journal of Science* 97:1–2.

Pinker, S. 1994. *The language instinct.* New York: HarperCollins.

Portmann, A. 1990. *A zoologist looks at humankind*. New York: Columbia University Press.

Potts, R. 1984. Home bases and early hominids. *American Scientist* 72:338–347.

Prahl-Andersen, B., and F. P. G. M. van der Linden. 1972. The estimation of dental age. *Transactions of the European Orthodontic Society* 48:535–541.

Premack, D. 1976. *Intelligence in ape and man*. Hillsdale, NJ: Erlbaum.

Premack, D. 1988. "Does the chimpanzee have a theory of mind?" revisited. In *Machiavellian intelligence: Social expertise and the evolution of intellect in monkeys, apes, and humans*, ed. R. Byrne and A. Whiten, 160–178. Oxford, UK: Oxford University Press.

Premack, D., and G. Woodruff. 1978. Does the chimpanzee have a theory of mind? *Behavioral and Brain Sciences* 4:515–526.

Rak, Y., W. H. Kimbel, et al. 1994. A Neanderthal infant from Amud Cave, Israel. *Journal of Human Evolution* 26:313–324.

Richerson, P., and R. Boyd. 2000. Climate, culture, and the evolution of cognition. In *Evolution of cognition*, ed. C. Heyes and L. Huber, 329–346. Cambridge, MA: MIT Press.

Richerson, P., and R. Boyd. 2003. Cultural evolution of human cooperation. In *Genetic and cultural evolution*, ed. F. Hammerstein, 373–404. Cambridge, MA: MIT Press.

Richmond, B. G. 2000. Evidence that humans evolved from a knuckle-walking ancestor. *Nature* 404:362–385.

Richmond, B. G., D. R. Begun, et al. 2001. Origin of human bipedalism: The knuckle-walking hypothesis revisited. *Yearbook of Physical Anthropology* 44:70–105.

Richmond B. G., and D. S. Strait. 2000. Evidence that humans evolved from a knuckle-walking ancestor. *Nature* 404:382–385.

Rightmire, G. P. 1979. Cranial remains of Homo erectus from Beds II and IV, Olduvai Gorge, Tanzania. *American Journal of Physical Anthropology* 51:99–115.

Rightmire, G. P. 1990. *The evolution of* Homo erectus: *Comparative anatomical studies of an extinct human species*. Cambridge, UK: Cambridge University Press.

Rose, M. 1991. The process of bipedalization in hominids. In *Bipedie chez les hominides*, ed. Y. Coppens and B. Senut, 37–48. Paris: CNRS Cahiers de Paleoanthropologie.

Ross, C. 1991. Life history pattern of New World monkeys. *International Journal of Primatology* 12:481–502.

Rowlett, R. M. 1999. Commentary on "The raw and the stolen." *Current Anthropology* 40:584–586.

Russon, A. 1996. Imitation in everyday use: Matching and rehearsal in the spontaneous imitation of rehabilitant orangutans (*Pongo pygmaeus*). In *Reaching into thought: The minds of great apes*, ed. A. Russon, K. Bard, and S. T. Parker, 152–176. Cambridge, UK: Cambridge University Press.

Russon, A. 1999. Imitation of tool use in orangutans. In *The mentalities of gorillas and orangutans*, ed. S. T. Parker, H. L. Miles, and R. W. Mitchell, 117–146. Cambridge, UK: Cambridge University Press.

Russon, A., K. Bard, et al., eds. 1996. In *Reaching into thought: The minds of great apes*. Cambridge: Cambridge University Press.

Russon, A. E., and D. R. Begun, eds. 2004. *The evolution of thought: Evolutionary origins of great ape intelligence.* Cambridge, UK: Cambridge University Press.

Russon, A. E., and B. Galdikas. 1993. Imitation in free-ranging rehabilitant orangutans (*Pongo pygmaeus*). *Journal of Comparative Psychology* 107 (2): 147–161.

Russon, A. E., and B. Galdikas. 1995. Constraints on great ape imitation: Model and action selectivity in rehabilitant orangutans (*Pongo pygmaeus*) imitation. *Journal of Comparative Psychology* 109:5–17.

Ruvolo, M. 1994. Molecular evolutionary processes and conflicting gene trees: The hominoid case. *American Journal of Physical Anthropology* 94:89–113.

Sacher, G. A. 1959. Relation of lifespan to brain weight and body weight in mammals. In *The lifespan of mammals*, ed. G. Wolstenhome and M. O'Connor. London: Churchill, CIBA Foundation colloquia on aging 5:115–133.

Sacher, G. A. 1975. Maturation and longevity in relation to cranial capacity in hominid evolution. In *Primate functional morphology and evolution*, ed. R. Tuttle, 417–441. The Hague: Mouton and Co.

Sacher, G. A. 1982. Relation of lifespan to brain weight and body weight in mammals. In *Primate brain evolution: Methods and concepts*, ed. E. Armstrong and D. Falk, 115–33. New York: Plenum.

Sacher, G. A., and E. Staffeldt. 1974. Relation of gestation time to brain weight for placental mammals. *American Naturalist* 108:593–615.

Sarich, V. 1968. Immunological time scale for hominoid evolution. *Science* 158:1200.

Sarich, V., and J. E. Cronin. 1976. Molecular systematics of the primates. In *Molecular anthropology*, ed. M. Goodman and R. E. Tashian, 141–170. New York: Plenum.

Savage-Rumbaugh, E. S., J. Murphy, et al. 1993. *Language comprehension in ape and child.* Chicago: University of Chicago Press.

Savage-Rumbaugh, E. S., M. A. Romski, et al. 1989. Symbol acquisition and use by *Pan troglodytes*, *Pan paniscus*, and *Homo sapiens*. In *Understanding chimpanzees*, ed. P. G. Heltne and L. A. Marquardt, 266–295. Cambridge, MA: Harvard University Press.

Savage-Rumbaugh, E. S., B. Wilkerson, et al. 1977. Spontaneous gestural communication among conspecifics in pygmy chimpanzees. In *Progress in ape research*, ed. G. Bourne, 287–309. New York: Academic Press.

Savage-Rumbaugh, S. 1995. *Symbolic communication in bonobos*. Jean Piaget Society, Berkeley CA: Erlbaum.

Savage-Rumbaugh, S., D. Rumbaugh, et al. 1978. Symbolic communication between two chimpanzees. *Science* 201:641–644.

Schaller, G., and G. Lowther. 1969. The relevance of carnivore behavior to the study of early hominids. *Southwest Journal of Anthropology* 25:307–341.

Scheib, J. E., S. W. Gangestad, et al. 1999. Facial attractiveness, symmetry, and cues of good genes. *Proceedings of the Royal Society of London: Series B* 266:1913–1917.

Schick, K., and N. Toth. 1993. *Making silent stones speak: Human evolution and the dawn of technology*. New York: Simon and Schuster.

Schick, K., and N. Toth. 1994. Early Stone Age technology in Africa: A review and case study into the nature and function of spheroids and subspheroids. In *Integrative paths to the past: Paleoanthropological advances in honor of F. Clark Howell*, ed. R. S. Corruccini and R. L. Ciochon, 429–449. Englewood Cliffs, NJ: Prentice Hall.

Schoetensack, O. 1908. *The mandible of* Homo heidelbergensis *from the Mauer sands at Heidelberg*. Leipzig: Wilhelm Engelmann.

Schultz, A. 1969. *The life of primates*. New York: Universe Books.

Schwartz, G. T., P. Mahoney, et al. 2005. Dental development in *Megaladapis edwardsi* (Primates, Lemuriformes): Implications for understanding life history variation in sub-fossil lemurs. *Journal of Human Evolution* 49:702–721.

Selander, R. K. 1972. Sexual selection and dimorphism in birds. In *Sexual selection and the descent of man, 1871–1971*, ed. B. Campbell, 180–230. Chicago: Aldine.

Semaw, S., P. Renne, et al. 1997. 2.5-million-year-old stone tools from Gona, Ethiopia. *Nature* 385:333–336.

Senghas, A., S. Kita, et al. 2004. Children creating core properties of language: Evidence from an emerging sign language in Nicaragua. *Science* 305 (5691): 1779–1782.

Senut, B., M. Pickford, et al. 2001. The first hominid from the Miocene (Lukeino formation, Kenya). *Comptes Rendus de l'Academie des Sciences, Paris* 332:137–144.

Shea, B. T. 1984. An allometric perspective on the morphological and evolutionary relationships between pygmy (*Pan paniscus*) and common (*Pan troglodytes*) chimpanzees. In *The pygmy chimpanzee*, ed. R. S. Susman, 89–130. New York: Plenum.

Shea, B. T., and S. F. Inouye. 1993. Knuckle-walking ancestors. *Science* 259:293–294.

Shingehara, N. 1980. Epiphyseal union, tooth eruption, and sexual maturation in the common tree shrew, with reference to its systematic problem. *Primates* 21:1–19.

Shipman, P. 1986. Scavenging or hunting in early Hominidae: Theoretical framework and tests. *American Anthropologist* 88:27–43.

Shipman, P. 2001. What can you do with a bone fragment? *Proceedings of the National Academy of Sciences* 98:1335–1337.

Short, R. V. 1981. Sexual selection in man and the great apes. In *Reproductive biology of the great apes*, ed. C. E. Graham, 319–341. New York: Academic Press.

Sillén-Tullberg, B., and A. P. Møller. 1993. The relationship between concealed ovulation and mating systems in anthropoid primates: A phylogenetic analysis. *The American Naturalist* 141:1–25.

Singh, D. 1993. Adaptive significance of female physical attractiveness: Role of waist-to-hip ratio. *Journal of Personality and Social Psychology* 65:293–307.

Singh, D., and P. M. Bronstad. 2001. Female body odour is a potential cue to ovulation. *Proceedings of the Royal Society of London: B* 268:797–801.

Small, M. F. 1995. *What's love got to do with it? The evolution of human mating.* New York: Doubleday.

Smith, B. H. 1986. Dental development in *Australopithecus* and early *Homo*. *Nature* 323:327–329.

Smith, B. H. 1989. Dental development as a measure of life history in primates. *Evolution* 43 (3): 683–688.

Smith, B. H. 1991a. Dental development and the evolution of life history in hominidae. *American Journal of Physical Anthropology* 86:157–174.

Smith, B. H. 1991b. Standards of human tooth formation and dental age assessment. In *Advances in dental anthropology*, ed. M. A. Kelley and C. S. Larsen, 143–168. New York: Wiley-Liss, Inc.

Smith, B. H. 1993. The physiological age of KNM-WT 15000. In *The Nariokotome Homo erectus skeleton*, ed. A. Walker and R. Leakey, 195–220. Cambridge, MA: Harvard University Press.

Smith, B. H. 1994. Patterns of dental development in *Homo*, *Australopithecus*, *Pan*, and *Gorilla*. *American Journal of Physical Anthropology* 94:307–325.

Smith, B. H., T. L. Crummett, et al. 1994. Ages of eruption of primate teeth: A compendium for aging individuals and comparing life histories. *Yearbook of Physical Anthropology* 37:177–231.

Smith, B. H., and R. L. Tompkins. 1995. Toward a life history of the hominidae. *Annual Review of Anthropology* 24:257–279.

REFERENCES

Smith, R. L. 1984. Human sperm competition. In *Sperm competition and the evolution of animal mating systems*, ed. R. L. Smith, 601–659. New York: Academic Press.

Spencer, F. 1984. The Neandertals and their evolutionary significance: A brief historical survey. In *The origins of modern humans: A world survey of the fossil evidence*, ed. F. H. Smith and F. Spencer, 1–49. New York: Alan R. Liss.

Sponheimer, M., and J. A. Lee-Thorp. 2003. Differential resource utilization by extant great apes and australopithecines. *Comparative Biochemistry and Physiology* 136:27–34.

Sponheimer, M., and J. Lee-Thorp. 1999. Isotopic evidence for the diet of an early hominid, *Australopithecus africanus*. *Science* 283:368–370.

Sponheimer, M., D. de Ruiter, et al. 2005. Sr/Ca and early hominin diets revisited: new data from modern and fossil tooth enamel. *Journal of Human Evolution* 48:147–156.

Sponheimer, M., J. A. Lee-Thorp, et al. 2005. Hominins, sedges, and termites: New carbon isotope data from the Sterkfontein Valley and Kruger National Park. *Journal of Human Evolution* 48:301–312.

Spoor, F., M. G. Leakey, et al. 2007. Implications of new early *Homo* fossils from Ileret, east of Lake Turkana, Kenya. *Nature* 448:688–691.

Spoor, F., B. A. Wood, et al. 1994. Implications of early hominid labyrinthine morphology for the evolution of human bipedal locomotion. *Nature* 369:645–648.

Stanford, C. 1996. The hunting ecology of wild chimpanzees: Implications for the behavioral ecology of Pliocene hominids. *American Anthropologist* 98:1–18.

Stanford, C. 1998. *Chimpanzees and red colobus: The ecology of predator and prey*. Cambridge, MA: Harvard University Press.

Stanford, C. 1999. *The hunting ape*. Princeton, NJ: Princeton University Press.

Stern, J. T., and R. L. Susman. 1983. The locomotor anatomy of *Australopithecus afarensis*. *American Journal of Physical Anthropology* 60:279–317.

Steward, J. 1955. *Theory of culture change*. Urbana: University of Illinois Press.

Stewart, C. B., and T. Disotell. 1998. Eurasian apes. *Current Biology* 8 (30 July/15Aug): R582–R588.

Stewart, T. D. 1977. The Neanderthal skeletal remains from Shanidar Cave, Iraq: A summary of findings to date. *Proceedings of the American Philosophical Society* 121:121–165.

Stiner, M. C. 2001. Thirty years on the "broad spectrum revolution" and Paleolithic demography. *Proceedings of the National Academy of Sciences of the USA* 98:6993–6996.

Stocking, J. G. W. 1968. *Race, culture, and evolution: Essays on the history of anthropology*. New York: Free Press.

Stoczkowski, W. 2002. *Explaining human origins: Myth, imagination and conjecture*. Cambridge, UK: Cambridge University Press.

Stoddart, D. M. 1990. *The scented ape: The biology and culture of human odour*. New York: Cambridge University Press.

Stokstad, E. 2000. Hominid ancestors may have knuckle walked. *Science* 287:2131–2132.

Stringer, C. B. 1984. Out of Africa—a personal history. In *Origins of anatomically modern humans*, ed. M. H. Nitecki and D. V. Nitecki, 150–174. New York: Plenum.

Stringer, C. B. 1989. The origin of early modern humans: A comparison of the European and non-European evidence. In *The human revolution: Behavioural and biological perspectives on the origins of modern humans*, ed. P. Mellars and C. B. Stringer, 232–244. Edinburgh: Edinburgh University Press.

Stringer, C. B. 1996. Current issues in modern human origins. In *Contemporary issues in human evolution*, ed. W. E. Meickle, F. C. Howell, and N. G. Jablonski, 115–134. San Francisco: California Academy of Sciences.

Stringer, C., and C. Gamble. 1993. *In search of the Neanderthals*. London: Thames and Hudson.

Stringer, C. B., J.-J. Hublin, et al. 1984. The origin of anatomically modern humans in western Europe. In *The origins of modern humans: A world survey of the fossil evidence*, ed. F. H. Smith and F. Spencer, 51–135. New York: Alan R. Liss.

Strum, S. C., D. Forster, et al. 1997. Why Machiavellian intelligence may not be Machiavellian. In *Machiavellian Intelligence II*, ed. A. Whiten and R. Byrne, 50–85. Cambridge, UK: Cambridge University Press.

Susman, R. 1988. Hand of *Paranthropus robustus* from Member 1, Swartkrans: Fossil evidence for tool behavior. *Science* 240:781–784.

Susman, R. L. 1998. Hand function and tool behavior in early hominids. *Journal of Human Evolution* 35 (1): 23–46.

Susman, R. L., J. T. Stern, et al. 1984. Arboreality and bipedality in the Hadar hominids. *Folia Primatologica* 43:113–156.

Susman, R. L., J. T. Stern, et al. 1985. Locomotor adaptations in the Hadar hominids. In *Ancestors: The hard evidence*, ed. E. Delson, 184–192. New York: Alan R. Liss.

Swaddle, J. P., and I. C. Cuthill. 1995. Asymmetry and human facial attractiveness: Symmetry may not always be beautiful. *Proceedings of the Royal Society of London: B* 261:111–116.

Swisher, C. C. I., W. J. Rink, et al. 1996. Latest *Homo erectus* of Java: Potential contemporaneity with *Homo sapiens* in Southeast Asia. *Science* 274:1870–1874.

Swisher, C. C., G. H. Curtis, et al. 1994. Age of the earliest known hominids in Java, Indonesia. *Science* 263:1118–1121.

Symons, D. 1979. *The evolution of human sexuality*. New York: Oxford University Press.

Szalay, F. 1976. Hunting-scavenging protohominids: A mode for hominid origins. *Man* 10:420–429.

Tanner, J. E., and R. W. Byrne. 1996. Representation of action through iconic gesture in a captive lowland gorilla. *Current Anthropology* 37 (1): 162–173.

Tanner, J. E., and R. W. Byrne. 1999. Spontaneous gestural communication in captive lowland gorillas. In *Mentalities of gorillas and orangutans*, ed. S. T. Parker, R. W. Mitchell, and H. L. Miles, 211–239. Cambridge, UK: Cambridge University Press.

Tanner, N. 1987. Gathering by females: The chimpanzee model revisited and the gathering hypothesis. In *The evolution of human behavior: Primate models*, ed. W. Kinzey, 3–27. Albany: State University of New York Press.

Tanner, N., and A. Zihlman. 1976. Women in evolution, part I: Innovation and selection in human origins. *Signs: Journal of Women in Culture and Society* 1:585–608.

Tattersall, I. 1986. Species recognition in human paleontology. *Journal of Human Evolution* 15:165–175.

Tattersall, I. 1995. *The last Neanderthal: The rise, success, and mysterious extinction of our closest human relatives*. New York: Macmillan.

Teleki, G. 1973. *The predatory behavior of wild chimpanzees*. Lewisburg, PA: Bucknell University Press.

Teleki, G. 1974. Chimpanzee subsistence technology: Materials and skills. *Journal of Human Evolution* 3:575–594.

Teleki, G. 1975. Primate subsistence patterns: Collector-predators and gatherer hunters. *Journal of Human Evolution* 4:125–184.

Terrace, H., L. Petitito, et al. 1979. Can an ape create a sentence? *Science* 206:891–902.

Thierry, B. 2005. Hair grows to be cut. *Evolutionary Anthropology* 14:5.

Thompson, D. D., and E. Trinkaus. 1981. Age determination for the Shanidar 3 Neanderthal. *Science* 212:575–577.

Thompson, M. E. 2005. Reproductive endocrinology of wild female chimpanzees (*Pan troglodytes schweinfurthii*): Methodological considerations and the role of hormones in sex and conception. *American Journal of Primatology* 67:137–158.

Thornhill, R., and S. W. Gangestad. 1994. Fluctuating asymmetry and human sexual behavior. *Psychological Science* 5:293–302.

Tinbergen, N. 1963. On aims and methods of ethology. *Zeitschrift fur Tierpsychologie* 20:410–429.

Tobias, P. V. 1987. The brain of *Homo habilis*. *Journal of Human Evolution* 16:741–761.

Tomasello, M., A. C. Kruger, et al. 1993. Cultural learning. *Behavioral and Brain Sciences* 16:495–552.

Tooby, J., and I. DeVore. 1987. The reconstruction of hominid behavioral evolution through strategic modeling. *The evolution of human behavior: Primate models*, ed. W. G. Kinzey, 183–287. Albany: State University of New York Press.

Toth, N., and K. Schick. 1993. Early stone industries and inferences regarding language and cognition. In *Tools, language and cognition in human evolution*, ed. K. R. Gibson and T. Ingold, 346–362. New York: Cambridge University Press.

Trinkaus, E. 1983. *The Shanidar Neanderthals*. New York: Academic Press.

Trinkaus, E. 1995. Neanderthal mortality patterns. *Journal of Archaeological Science* 22:121–142.

Trinkaus, E., and D. D. Thompson. 1987. Femoral diaphyseal histomorphometric age determinations for the Shanidar 3, 4, 5, and 6 Neandertals and Neandertal longevity. *American Journal of Physical Anthropology* 72:123–129.

Trivers, R. 1971. The evolution of reciprocal altruism. *Quarterly Review of Biology* 46:35–57.

Trivers, R. 1972. Parental investment and sexual selection. In *Sexual selection and the descent of man*, ed. B. Campbell, 136–179. New York: Aldine.

Tuttle, R. 1969. Knuckle-walking and the problem of human origins. *Science* 166:953–961.

Tuttle, R., B. Hallgrimsson, et al. 1999. Electromyography, elastic energy, and knuckle walking: A lesson in experimental anthropology. In *The new physical anthropology*, ed. S. S. Strum, D. Lindberg, and D. Hamburg, 32–45. Upper Saddle River, NJ: Prentice-Hall, Inc.

Tuttle, R., D. M. Webb, et al., eds. 1991. *Laetoli footprint trails and the evolution of hominid bipedalism*. Cahiers de Paleoanthropologies. Paris: Editions du CNRS.

Underhill, P. A., P. Shen, et al. 2000. Y chromosome sequence variation and the history of human populations. *Nature Genetics* 26:358–361.

Van der Merwe, N. J., J. F. Thackeray, et al. 2003. The carbon isotope ecology and diet of *Australopithecus africanus* at Sterkfontein, South Africa. *J Hum Evol* 44 (5): 581–597.

Van Hooff, J. A. R. A. M. 1972. A comparative approach to the phylogeny of laughter and smiling. In *Nonverbal communication*, ed. R. A. Hinde, 209–237. Cambridge: Cambridge University Press.

Van Hooff, J. A. R. A. M., and S. Preuschoft. 2003. Laughter and smiling: The intertwining of nature and culture. In *Animal social complexity*, ed. F. de Waal and P. Tyack, 260–287. Cambridge, MA: Harvard University Press.

Van Lawick-Goodall, J. 1970. Tool-using in primates and other vertebrates. *Advances in the study of behavior*, vol. 3, ed. D. S. Lehrman, R. A. Hinde, and E. Shaw, 195–249. New York: Academic Press.

Van Peer, P. 1991. Interassemblage variability and Levallois styles: The case of a northern African Middle Paleolithic. *Journal of Anthropological Archaeology* 10:107–151.

Van Schaik, C., M. Ancrenaz, et al. 2003. Orangutan cultures and the evolution of material culture. *Science* 299 (5603): 102–105.

Van Schaik, C., R. Deaner, et al. 1999. The conditions for tool use in primates: Implications for the evolution of material culture. *Journal of Human Evolution* 36:719–741.

Van Schaik, C. P., E. A. Fox, et al. 1996. Manufacture and use of tools in wild Sumatran orangutans: Implications for human evolution. *Naturwissenschaften* 83:186–188.

Van Schaik, C. P., and C. D. Knott. 2001. Geographic variation in tool use on Neesia fruits in orangutans. *American Journal of Physical Anthropology* 114:331–342.

Vigilant, L., M. Stoneking, et al. 1991. African populations and the evolution of human mitochondrial DNA. *Science* 253:1503–1507.

Visalberghi, E., and D. Fragaszy. 1990. Do monkeys ape? In *"Language" and intelligence in monkeys and apes*, ed. S. T. Parker and K. R. Gibson, 247–273. New York: Cambridge University Press.

Visalberghi, E., and D. Fragaszy. 2002. Do monkeys ape? Ten years after. In *Imitation in animals and artifacts*, ed. K. Dautenhahn and C. Nehaniv, 471–499. Cambridge, MA: MIT Press.

Vogel, J. O., ed. 1997. *Encyclopedia of precolonial Africa*. Walnut Creek, CA: AltaMira Press.

Walker, A., and R. Leakey, eds. 1993. *The Nariokotome Homo erectus skeleton*. New York: Cambridge University Press.

Walker, A., and P. Shipman. 1996. *The wisdom of bones: In search of human origins*. Guernsey, Channel Islands: The Guernsey Islands Press.

Washburn, S. L. 1960. Tools and human evolution. *Scientific American* 203 (3): 62–75.

Washburn, S. L. 1967. Behavior and the origin of man. *Proceedings of the Anthropological Institute of Great Britain & Ireland* 3:21–27.

Washburn, S. L., and C. S. Lancaster. 1968. The evolution of hunting. In *Man the hunter*, ed. R. B. Lee, 293–303. Chicago: Aldine.

Watts, D. P., and A. E. Pusey. 1993. Behavior of juvenile and adolescent great apes. In *Juvenile primates*, ed. M. E. Pereira and L. A. Fairbanks, 148–167. New York: Oxford University Press.

Weidenreich, F. 1939. On the earliest representatives of modern mankind recovered on the soil of East Asia. *Bulletin of the Natural History Society of Peking* 13:161–174.

Wells, S. 2002. *The journey of man*. New York: Random House.

Westcott, R. 1967. Hominid uprightness and primate display. *American Anthropologist* 69:738.

West-Eberhard, M. J. 1983. Sexual selection, social competition, and speciation. *Quarterly Review of Biology* 58 (2): 155–183.

Whallon, R. 1989. Elements of culture change in the later Paleolithic. In *The human revolution*, ed. P. Mellars and C. Stringer, 433–454. Princeton, NJ: Princeton University Press.

Wheeler, P. F. 1984. The evolution of bipedality and functional body hair in hominids. *Journal of Human Evolution* 13:91–98.

Wheeler, P. 1991. The influence of bipedalism on the energy and water budgets of early hominids. *Journal of Human Evolution* 21:117–136.

Wheeler, P. 1992. The influence of the loss of functional body hair on hominid energy and water budgets. *Journal of Human Evolution* 23:379–388.

White, L. A. 1959. *The evolution of culture*. New York: McGraw Hill.

White, R. 1989. Production complexity and standardization in early Aurignacian bead and pendant manufacture: Evolutionary implications. In *The human revolution*, ed. P. Mellars and C. Stringer, 366–390. Princeton, NJ: Princeton University Press.

White, T., B. Asfaw, et al. 2003. Pleistocene *Homo sapiens* from Middle Awash, Ethiopia. *Nature* 423:742–747.

White, T. D., G. Suwa, et al. 1994. *Australopithecus ramidus*, a new species of early hominid from Aramis, Ethiopia. *Nature* 371:306–312.

Whiten, A., and D. Custance. 1996. Studies of imitation in chimpanzees and children. In *Social learning in animals: The roots of culture*, ed. B. G. Galef and C. M. Heyes, 291–318. New York: Academic Press.

Whiten, A., and R. Byrne. 1986. The St. Andrews catalogue of tactical deception in primates. *St. Andrews Psychological Report*, no. 10. St. Andrews, UK: St. Andrews University.

Whiten, A., and R. Byrne, eds. 1997. *Machiavellian intelligence II*. Cambridge, UK: Cambridge University Press.

Whiten, A., J. Goodall, et al. 1999. Cultures in chimpanzees. *Nature* 399:682–685.

Wiley, E. O. 1981. *Phylogenetics: The theory and practice of phylogenetic systematics*. New York: John Wiley.

Williams, G. 1966. *Adaptation and natural selection*. Princeton, NJ: Princeton University Press.

Williams, G. C. 1992. *Natural selection: Domains, levels, and challenges*. New York: Oxford University Press.

Willoughby, D. P. 1978. *All about gorillas*. Cranbury, NJ: A.S. Barnes and Co., Inc.

Wolde-Gabriel, G., Y. Haile-Selassie, et al. 2001. Geology and paleontology of the Late Miocene Middle Awash Valley, Afar rift, Ethiopia. *Nature* 412:175–178.

Wolpoff, M. H., J. D. Hawks, et al. 2001. Modern human ancestry at the peripheries: A test of the replacement theory. *Science* 291:293–297.

Wolpoff, M. H., A. G. Thorne, et al. 1984. Multiregional evolution: A worldwide source for modern human populations. In *Origins of anatomically modern humans*, ed. H. Nitecki and D. V. Nitecki, 175–199. New York: Plenum.

Wolpoff, M. H., W. X. Zhi, et al. 1984. Modern *Homo sapiens* origins: A general theory of hominid evolution involving the fossil evidence from East Asia. In *The origins of modern humans: A world survey of the fossil evidence*, ed. F. H. Smith and F. Spence, 411–483. New York: Alan R. Liss.

Wood, B. A. 1991. *Koobi Fora research project IV: Hominid cranial remains from Koobi Fora*. Oxford: Clarendon.

Wood, B. 1992. Origins and evolution of the Genus *Homo*. *Nature* 355:783–790.

Woodruff, G., and D. Premack. 1979. Intentional communication in the chimpanzee: The development of deception. *Cognition* 7:333–362.

Wrangham, R. 1987. The significance of African apes for reconstructing human social evolution. In *The evolution of human behavior: Primate models*, ed. W. G. Kinzey, 51–71. Albany: State University of New York Press.

Wrangham, R. 1999. Evolution of coalitionary killing. *Yearbook of Physical Anthropology* 42:1–30.

Wrangham, R. W. 2001. Out of the pan, into the fire: How our ancestors' evolution depended on what they ate. In *Tree of origin: What primate behavior can tell us about human social evolution*, ed. F. de Waal, 121–172. Cambridge, MA: Harvard University Press.

Wrangham, R., J. H. Jones, et al. 1999. The raw and the stolen. *Current Anthropology* 40 (5): 567–594.

Wrangham, R., and D. Peterson. 1996. *Demonic males: Apes and the origins of human violence*. Boston: Houghton Mifflin.

Wynn, T. 1979. The intelligence of later Acheulean hominids. *Man* 14:171–191.

Wynn, T. 1989. *The evolution of spatial competence*. Urbana: University of Illinois Press.

Wynn, T., and W. McGrew. 1989. An ape's view of the Oldowan. *Man* 24:388–397.

Yerkes, R. M., and A. Yerkes. 1929. *The great apes*. New Haven, CT: Yale University Press.

Young, R. W. 2003. Evolution of the human hand: The role of throwing and clubbing. *Journal of Anatomy* 202 (1): 165–174.

Zahavi, A., and A. Zahavi. 1997. *The handicap principle*. New York: Oxford University Press.

Zeresenay, A., F. Spoor, et al. 2006. A juvenile early hominin skeleton from Dikika, Ethiopia. *Nature* 443:296–301.

Zhisheng, A., and H. C. Kun. 1989. New magnetostratigraphic dates of Lantian *Homo erectus*. *Quaternary Research* 32:213–221.

Zihlman, A. 1978. Women in evolution, part II: Subsistence and social organization among early hominids. Signs: *Journal of Women in Culture and Society* 4:4–20.

Zihlman, A. 1999. Fashions and models in human evolution: Contributions of Sherwood Washburn. In *The new physical anthropology*, ed. S. S. Strum, D. G. Lindberg, and D. Hamburg, 151–161. Upper Saddle River, NJ: Prentice-Hall.

Zihlman, A., J. Cronin, et al. 1978. Pygmy chimpanzees as possible prototype for the common ancestor of humans, chimpanzees and gorillas. *Nature* 275:744–746.

Zihlman, A., and N. Tanner. 1979. Gathering and hominid adaptation. In *Female hierarchies*, ed. L. Tier and H. Fowler, 163–194. Chicago: Beresford Book Services.

Index

Page numbers in italics indicate figures or tables.

adaptations, 5, 6, 8, 11, 14, 15, 22–24, 26, 38, 46, 48, 58. 60, 61, 62, 63, 64, 66, 69, 70, 75, 77, 80, 82, 88, 89, 93, 112, 113, 125, 133, 149–152, 158, 159, 163, 167, 168, 174, 185, 197, 200–202

adaptive criteria: correlational, *23*, 102, 187; phyletic, 22, *23*, 27, 66, 79

adaptive radiations, 22, 42, 66

African apes: Bonobos, 20, *21*, 24, 26, 27, 45, 46, 71, 72, 126, 127, 133, 137, 138, 146, 147, 176–177, 181, 205, 207, 208; chimpanzees, 4, 9, 11, 20, *21*, 22–27, 34, 36, 45–49, 55–57, 63, 65, 70–72, 74, 75, 80, 87, 89, 91, 92, 96, 99, 100, 110, 114–116, 122, 126, 127, 133, 136–139, 141, 146,147, 156–161, 167, 170, 172–173, 180–182, 186, 188, 190, 192, 202, 205, 207, 208; gorillas, 20, *21*, 22, 25, 27, 71, 74, 87, 89, 109, 120, 127, 133, 137, 138, 176–177, 181, 205

Aiello, L., 92, 98, 104, 194, 202, 203

aimed throwing, 67, 77, 78, 80–82, 120, 150, 158, 159, 163, 165, 172, 200, 202

Armstrong, D., 140, 141, 143, 151

apomorphy. *See* character state: derived

apprenticeship, 148, 159, 166, 168, 172, 183, 190, 191, 20

Ascher, S., 134, 135, 143

assessment, 39, 81, 91, 106, 114, 117, 124, 144, 149, 164, 170, 179, 180

Australopithecus, 54, 58, 62, 72, 84, 86, 111, 139, 146, 170, 182

Australopithecus-like species: *Paraustralopithecus aethiopicus*, *33*, 36; *Paranthropus robustus*, 31, *33*, 36, 49, 78, 93; *Zinjanthropus (Paranthropus) boisei*, *33*, 36, 93

Australopithecus species: *A. afarensis*, 31, *33*, 36, 72, 74, 75, 81, 83, 88, 93, 113; *A. africanus*, 31, *33*, 36, 50, 83, 92, 93; *A. anamensis*, 31, *33*, 36; *A. bahrelghaszli*, 31, *33*, 36; *A. garhi*, *31*, *33*, 36, 38, 140, 165, 190

243

autapomorphy. *See* character states: uniquely derived

Bartholomew, G., 46, 47, 55, 63, 79
basal hominins, 30, 32, 34, 48, 49, 63, 65, 71, 72, 76, 92, 140, 163, 190
basal hominin species: *Ardipithecus ramidus*, 30, *33*, 35, 72; *Orrorin tugenesis*, 30, *33*, 35, 72; *Sahelanthropus tchadensis*, 30, *33*, 35, 72
beauty, 129, 130, 155
Bigelow, R., 192, 193, 202, 204
Bingham, P., 77, 80, 200–203
bipedalism scenarios: carrying hypothesis, 57, 67, 72, 73, 77, 79, 80, 82, 89, 134, 146; upright display hypothesis, 77, 128; throwing hypothesis, 73, 77, 78, 79, 80–82, 120, 165, 172, 183, 200–202
Birdsell, J., 46, 47, 55, 63, 73, 79
Blumenschine, R., 58, 60
body display scenarios: aquatic ape hypothesis, 110, 112; sexual signals hypothesis, 118, 119, 130; temperature regulation hypothesis, 110, 112, 113, 128
Boesch, C., 11, 46–48, 59, 138, 160, 188, 190
bone development, 87
Bromage, T., 90, 96, 99
Brooks, A., 32, 39, 40, 48, 49, 51–53, 65, 129, 142, 163, 166–168, 192
Boyd, R., 53, 167, 168, 183, 194, 195
brain: energy costs, 97, 98; evolution, 155, 159, 180; size, 10, 21, 50, 76, 85, *87*, 97–99, 104, 125, 139, 141, 145, 147, 152, 156, 168, 171, 173, 178, 191, 192, 196

canine tooth reduction, 4, 6, 10, 26, 35, 36, 56–58, 61, 76, 81, 82, 88, 89, 134, 192, 205, 207
character states: ancestral, *16*, 18, 26, 31, 51, 55, 71, 73, 126, 147, 159, 176; derived, *16*, 206; shared ancestral, *16*, 17–20, 22, 205; shared derived, *16*, 17, 19, 20, 22, 23, 26, 32, 66, 79, 128, *206*, 207, 208; uniquely derived, *16*, 19, 20, 23, 24, 128, 205, 207
choice (intersexual): female, 56, 57, 107, *109*, 117, 152, 177, 178, 184, 191, 204; male, *109*, 124, 128
cladistics, 23, 27, 30
cladogram, 18, *19*, 20, 22, *29*
climate, 32, 53, 55, 100, 102, 128, 167, 168, 182, 195
cognition. *See* intelligence
communication: distal, 128, 149, 153; mother-infant, 146, 147; referential, 133, 141, 143, 148, 159, 172, 191; symbolic, 138, 183, 190, 192; vocal, 116, 128, 134, 136, 137, 140, 143, 145, 147, 151, 197
comparative studies 26, 27, 54, 60, 67, 71, 97, 128, 138, 156, 161, 203
competition (intrasexual): female, *109*, 128; male, *109*, 117, 149, 152, 177, 180, 183, 204; sperm, 109, 120, 122, 123, 127
confidence in paternity, 107, 126
continuity approach, 7, 79, 82, 141, 143, 147–149, 151, 183, 203
control of females, *109*, 176, 178
cooking, 58, 62, 63, 64, 166
cooperation: kin, 58, 153, 194, 203, 204; for conflict, 183, 192; nonkin, 194, 200, 202–204
culture: art, 143, 151, 167, 183, *189*, 198; change, 142, 193–97; chimpanzee, 188, 190, 196; definitions of, 185, 186; material, 89, 141, 142, 156, 161, 186, 191; morals, 5, 6, 7, 155, 177, 179, 180, 185, 192, 194, 201; music, 131, 142, 146, 149, 177–180, 183, 198, 204; origins of, *199*, 203; trade, 66, 186, 192, 193, 204
culture scenarios: feedback, 186, 195; remote killing, 200, 203, 204; warfare, 53, 152, 175, 180, 186, 192, 193, 202, 204

INDEX

Dart, R., 50, 51, 55, 58, 72, 77–79, 92
Darwin, C., 3, 4, 6, 7, 9–11, 14, 25, 41, 42, 50, 58, 67, 72, 76, 77, 80, 82, 106, 109, 110, 128, 129, 149, 155, 177, 178, 183, 185, 187, 192, 194, 195
Davidson, I., 139, 150, 151
Dean, C., 76, 90, 96, 99, 152
death, determining age at, 87, 90, 93, 95, 97, 100, 200
dental development, 87–93, 96–99, 104, 163
diet, 45, 48, 55, 56–59, 62, 73, 98, 104, 134
distributed knowledge, 195, 198, 204
DNA, 18, 19, 32, 41, 107, 213, 223, 239
Donald, M., 148, 196–201, 203, 204
Dunbar, R., 144, 145, 149

early human ancestors, 31, 35, 42, 49, 55, 64, 71–73, 76, 77, 82, 92, 93, 139, 165, 190, 194, 197
Engel, F., 76, 187, 195
environment, 40, 47, 50, 102, 136, 137, 150, 193, 195
evolution, 5, 7, 8, 10, 11, 13, 14, 16, 18, 20, 22, 23, 25–27, 30, 42, 46, 58, 81, 82, 85, 88, 116, 118
explanations: proximate, 8, 9, 132, 143, 148, 174, 179; ultimate, 179

faces: expressions, 26, 114, *115*, 116, 133; features, 106, 114, 117, 120, 153; cultural elaboration of, 129, 130
Falk, D., 146, 147, 149
female body displays: breasts, 109, 110, 118, 119, 122–124, 128, 130; hips, 109, 110, 119, 124, 128, 130
female reproduction: concealed estrus, 58, 89, 125, 127, 128, 182; lactation, 95, 103, 207; menopause, 85, 95, 100, 102, 124; orgasm, 125, 127, 179; sperm manipulation, 125
fire, 60, 62, 63, 64, 65, 76, 128, 141, 166, 187, 197, 201

fitness, 14, 101, 106, 107, 110, 149
food sharing, 4, 59, 94, 95, 100, 103, 104, 143, 159, 172, 179, 180
fossil hominin body sizes, *36*

geologic time periods: Miocene, 6, 25, 26, 32, 34, 42, 61, 88, 89, 94, 111, 134, 143, 158, 167; Plio-Pleistocene, 6, 26, 42, 60, 94, 167
Gibson, K. R., 23, 48, 49, 55, 66, 79, 148, 153, 157–160, 168, 173, 180, 182, 203
great apes. *See* African apes; orangutans
Guthrie, D., 112, 114, 117, 120, 121, 130

habitats, 9, 10, 42, 48, 49, 52, 56–58, 60, 61, 63–65, 74, 76, 89, 101, 142, 153, 171, 172, 174, 195
hair: body, 7, 106, 110–113, 118, 119, 128, 130; facial, 116, 120, 129, 130; head, 120, 130; loss, 68, 105, 110–113, 128, 133
handicap principle, 149, 177, 178, 180
hands: carrying, 57, 67, 72, 73, 77, 79, 80, 82, 89, 134, 146; climbing, 74; freed, 6, 57, 67, 72, 76, 141, 192; grips, 78, 80, 81, 137; throwing, 50, 80, 81, 201
Hardy, A., 79, 111, 112
Hawkes, K., 55, 58, 61, 85, 99, 100, 102, 104
Hewes, G., 11, 72, 73, 78, 79, 136, 137, 143, 153
Hockett, C., 134, 135, 143
hominidae (family), 27, 30, *36*, 207
homininea (subfamily), 27, 29, *36*, 30, 71
hominini (tribe), 30, 32, *36*
hominoidea (superfamily), 27, *36*, 207
Homo species: *H. antecessor*, *33*, 39; *H. erectus*, 9, 31, *33*, 37–39, 50, 54, 58–60, 82–83, 91, 96–104, 128, 129, 140, 141, 148, 152, 163, 166, 171–174, 182, 191, 196, 197, 199–201; *H. ergaster*, 31, *33*, 37–39, 83; *H. habilis*, 31, *33*, 36–38, 93, 99, 141, 148, 159, 172, 190; *H. heidelbergensis*, *33*, 39, 233; *H. neanderthalensis*, 31, *33*,

39, 40; *H. rudolfensis*, 33, 36, 37; *H. sapiens* (archaic), 10, 39, 41, 51, 52, 63, 65, 96, 97, 99, 142; *H. sapiens* (modern), 40, 41, 42, 51, 52, 104, 142, 144, 145, 166, 173, 191, 193

imitation, 7, 131, 137, 143–147, 148, 155–160, 166, 186, 188, 190, 194, 197, 205
immature hominin fossils: Dederiyah youngsters, 83; Dikika baby, 74, 83, 92; Taung child, 82, 83, 92; Turkana boy (aka Nariokotome boy) 38, 74, 83, 96
intelligence: great ape, 138, 151, 158, 161, 163, *169*, 196; Machiavellian intelligence, 71, 167; social, 116, 138, 145, 160, 166, 167, 170, 171, 173; spatial, 138, 163–165, 172
intelligence scenarios: extractive foraging hypothesis, 48, 55, 62–65, 82, 148, 153, 158, 163, 165, 168, 172; social intelligence hypothesis, 145, 166, 167, 171, 173, 181, 182

juvenile provisioning, 93, 95, 103
juvenilization, 207

Key, C. A., 98, 104, 202, 203

Lancaster, C., 54, 55, 58, 59, 79, 93–95, 103, 104, 203
Lancaster, J., 79, 93–95, 101, 103, 104, 203
Landau, M., 5, 13, 68
language: acquisition, 132, 146, 147, 149; centers (in brain), 139, 141; creole, 132; gestural, 136, 137, 140, 141, 143, 145, 151–153, 159; grammar, 132, 133, 140, 141, 147, 148; hyoid bone, 139–171; protolanguage, 131, 132, 136, 143, 146–150, 159; sign, 133, 136, 140, 145; spoken, 131, 132, 136–138, 140, 143; vocal tract, 132, 139, 141

language scenarios: gestural origins, 136, 137, 140, 141, 143, 145, 146, 151, 152, 159 153; motherese, 147, 149, 152; planning, 150–152, 160, 183; syntax, 140, 141, 143, 148, 152; vocal grooming, 144, 145, 149
last common ancestor (of humans and chimpanzees), 9, 11, 30, 32, 34, 41, 42, 46, 48, 57, 68, 82, 92, 126, 133, 143, 156, 157, 160, 190, 194, 205, 207
Leakey, R., 37, 38, 50, 74, 76, 77, 90, 91, 96, 97, 99
lesser apes, 20, 22, 26, 27, 69, 127, 134, 157, 205
life history configurations: altricial infancy, 85, 201; precocial infancy, 85; K-configuration, 84, 85; r-configuration, 84
life history scenarios: grandmothering hypothesis, 101, 104; pair bonding hypothesis, 118, 125, 129; provisioning young, 88, 93–95, 100–103, 118
life history stages: adolescence, 85, 95, 103, 177; childhood, 3, 85, 103; infancy (before weaning), 40, 85, *87*, 170, 173
life span (longevity), 61, 84, 85, *87*, 94, 97–102, 104; puberty, 7, 85, *87*, 90, 98, 99, 110, 114, 116, 120, 123, 124, 179
locomotion: bipedalism, 3, 4, 6, 7, 11, 22, 35, 36, 38, 57, 67, 68, 71–74, 76–80, 82, 88, 89, 93, 102, 113, 118, 120, 121, 128, 140, 141, 183, 192; brachiation, 15, 20, 22, 68–70, 74, 205; knuckle walking, 20, 22, 26, 50, 57, 67, 68, 70–72, 187, 205; quadrupedal, 69–71, 73, 74, 77, 113; tree climbing, 10, 36, 49, 57, 60, 68, 71, 72, 74, 76, 82, 134;
Lovejoy, O., 68, 79, 83, 88, 89, 101, 102, 104, 125, 126, 203

male body displays: bipedal, 77, 79, 81, 82, 117, 121; chest and shoulders, 6, 81, 110, 120, 128, 130, 201; facial, 120; olfactory, 112; penis, 109, 121, 122, 127, 128, 130; vocal threat, 120

Marzke, M., 77, 78, 80
McBrearty, S., 39, 40, 52, 53, 65, 129, 142, 167, 168, 192
McGrew, W., 172, 188, 190
McHenry, H., 36, 69, 91, 207
memory, 148, 196, 198–201
mentality, 155, 166, 170, 172, 174, 180, 181, 186, 203
mental representation, 143, 148, 152, 158, 161, 163, 197, *199*, 200, 201
middle period human ancestors, 128, 142
Miller, G., 126, 149, 174, 179, 181, 183, 204
Mithen, S., 139, 170, 171, 173, 174, 181, 182
modern human origins models: out of Africa, 41, 135, 142, 166, 173; regional continuity, 41
monkeys 7, 16–20, *21*, 23, 26, 47, 66, 85, 87, 127, 133, 139, 157, 168, 170, 190, 192
Morris, D., 110, 116, 118, 119,, 21, 123, 125, 126

Nariokotome boy. *See* immature hominin fossils
narratives, 4, 5, 8, 198–200
Neanderthals, 9, 40. 41, 52, 97, 100, 142, 143, 193, 202
neoteny. *See* juvenilization
niche construction, 195
Noble, W., 139, 150, 151

O'Connell, J., 55, 58, 61, 79, 99, 100–104, 203
orangutans, *19*, 20, 26, 27, *29*, 109, 133, 139, 157, 175, 181, 208

parental investment, 84, 87, 89, 93, 95, 98, 101, 102, 104, 107, 128
Parker, S. T., 23, 79, 120, 148, 149, 151, 159, 160, 168, 182, 203
phylogenetics, 15, 16, 22, 24, 127
Piagetian stages: concrete operations, 161, 163, 165, 166, 182, 196; formal, 161, 165, 173, 182; preoperations, 158, 161–165, 168, 172, 182, 190
plesiomorphy. *See* character states: ancestral

raiding. *See* warfare
recapitulation, 159, 170, 172, 173
reproductive strategies: pair bonding, 87, 89, 118, 125, 129; monogamy, 104; polyandry, 109; polygamy, 193; polygyny, 85, 109, 192
Richerson, P., 53, 55, 167, 168, 182, 194, 195
Richmond, B., 68, 69, 71, 72
ritualization, 105, 106, 216, 119, 121

self-awareness, 156, 168, 217, 219, 227, 229, 230
selection: group, 192, 194; kin selection, 11, 56, 153, 183, 200, 203; natural selection, 5–7, 13, 14, 45, 57, 78, 112, 155, 177, 178, 179, 182, 192, 195; reciprocal altruism, 182, 194; runaway, 106, 121, 177, 178
sexual dimorphism, 36, 47, 49, 61, 89, 91, 104, 110, *122*, 129, 134, 207
sexual selection, 6–8, 11, 22, 45, 56, 57, 79, 81, 105–107, 109–112, 120, 121, 127, 129, 144, 149, 174, 177–181, 192, 202, 204, 207
shelters, 140, 141, 166, 191
Shipman, P., 49, 56, 59, 60, 74
Stoczkowski, W., 3–5, 11, 13
Stokoe, W. C., 11, 140, 141, 143, 151
Strait, D. S., 71, 72
subsistence scenarios: excavating underground storage organs (USO), 4, 55, 56, 103, 159; environmental, 46, 51; extractive foraging, 4, 48, 55, 62, 63, 66, 82, 153, 157, 158, 168, 172, 173; gathering, 4, 46, 54–59, 79, 93, 95, 103; hunting, 4, 46–48, 51, 52, 54–56, 58, 59, 60, 62–65, 76–78, 82, 93, 100, 103, 111, 118, 129, 134, 137, 143, 152, 159, 166, 170, 183, 187, 191, 197; scavenging, 3, 11, 38, 48, 49, 55, 56, 58–61, 63, 64, 73, 76–80, 82, 103, 129, 134, 141, 165, 183, 191

symmetry, 110, 117, 164, 171, 173, 179
Symons, D., 125, 126
symplesiomorphy. *See* character states: shared ancestral
synapomorphy. *See* character states: shared derived
Szalay, F., 55, 56, 61

Tanner, N., 11, 30, 55–57, 79, 87, 138
taxonomic relevance, 17
taxonomies: cladistic, 27, 30; molecular, 10, 27, 41, 68, 72; old vs. revised, 27, *28*, 30; primate, *21*, 25, 26, *28*
tool technologies: Acheulean, 38, 50, 51, 54, 64, 76, 141, 163, 165, 166, 171–173, 186, 191, 192; Middle Stone Age, *26*, 51, 52, *53*, 129, 142, 156, 168, 192; Middle Paleolithic (Mousterian), 40, 52, 64, 65, 142, 162, 166, 191; Oldowan, 10, 38, 163–165, 170, 172, 173, 186, 190; Upper Paleolithic, 51, 52, 137, 142, 151, 171, 192, 193, 198, 201

tool use, 4, 11, 15, 23, 24, 26, 46–48, 57, 60, 72, 73, 76, 78, 79, 82, 136, 148, 156–157, 160, 163, 168, 170, 172, 208
Turkana boy. *See* immature fossils

underground storage organs (USO), 51, 55, 62, 64, 66, 101, 165, 166

waist-to-hip ratio, 119, 123, 124, 130
Walker, A., 37, 38, 59, 74
Washburn, S. L., 54, 55, 58, 59, 68, 72, 79, 175
Westcott, R,. 77, 121
Wheeler, P., 76, 110, 112, 113, 128
Wilcox, S., 11, 140, 141, 143, 151
Wrangham, R., 11, 30, 46, 55, 58, 62, 63, 78, 79, 174–176, 181, 183, 204, 208
Wynn, T., 163–165, 168, 170, 172, 173, 181

Young, R., 77, 80, 81

Zihlman, A., 11, 55–58, 79, 207

About the Authors

Sue T. Parker received her PhD at the University of California, Berkeley, in 1973. She is professor emeritus of anthropology at Sonoma State University, where she taught biological anthropology for thirty years. During her career she studied the development of human, group-living macaque, chimpanzee, and gorilla infants. She also worked as a Senior Research Fulbright Fellow at the Comparative Animal Behavior Laboratory in Rome in 1986. Her comparative research on the evolution of mental development in human and nonhuman primates can be found in numerous articles including those in six volumes she coedited on this topic, and in a previous book, *Origins of Intelligence in Monkeys, Apes, and Humans*, coauthored with Michael McKinney. She is a fellow of the California Academy of Sciences.

Karin E. Jaffe received her PhD at the University of California, Davis, in 2002. She is an associate professor in the department of anthropology and an adjunct faculty member in the department of biology at Sonoma State University, where she teaches courses in primate behavior, human evolution, and forensic anthropology. She is also program director of SSUPER: *S*onoma *S*tate *U*niversity *P*rimat*E R*esearch project. Dr. Jaffe's research on vervet (*Cercopithecus aethiops*) and patas (*Erythrocebus patas*) monkey antipredator behavior has appeared in several journals and edited volumes, including the *American Journal of Physical Anthropology*, *International Journal of Primatology*, *Folia Primatologica*, and *Primate Anti-Predator Strategies*, edited by S. Gursky and K. A. I. Nekaris.